Graven Image

Why The Earth Is Not A Globe?

Vincent Rhodes

Table of Contents

Introduction: Why Is The Globe A Graven Image? 1

God Wants You Separate From The World! 2

Babylonian gods 5

Chapter 1: The Hebrews View of the Earth 15

Chapter 2: The Greeks View of the Earth 36

Chapter 3: Men of the Globe 52

Chapter 4: The Cult of the Globe 74

Chapter 5: The Rotation of the Earth and Worship of the Sun 81

Chapter 6: How the Globe Supports A Fake Space Narrative? 90

INTRODUCTION
Why Is The Globe A Graven Image?

Exodus 20:4 says, "You shall not make for yourself any image or carving of anything in the Heavens above, the earth beneath or in the waters under the earth." God's love for humanity was evident after he created Adam and Eve. Adam and Eve rebelled against God in the garden of Eden, but God still loved them. Even after we all rebelled, he still pursued a relationship with us; so, he chose the Hebrews (Jews) to be his special people. Following the Hebrews, he chose the (gentiles) through his only begotten Son, Jesus Christ. It was unto the Hebrews that God revealed the Bible. The Hebrews are the primary writers of the Bible. It is through the writings of the Hebrews that we discover that the earth is not a globe. It is also through the writings of the Hebrews that we discover that: "God so loved the world that He gave His only begotten Son, that whosoever believes on him should not perish, but have everlasting life" (John 3:16). We gained eternal life through our Lord and Savior, Jesus Christ. He was crucified for our sins of rebellion against God, and his death on the cross and resurrection from the dead, paid our sin debt in full. God did this so that we (Jews & gentiles) would be his people. "You are a chosen generation, a royal priesthood, a holy nation, God's own people... he hath called you out of darkness into HIS marvelous light" (1 Peter 2:9). This means that we are separate from the world. We are to trust the Bible above worldly teachings. You were taught in school that the earth is a globe; however, the Bible disputes this claim. The globe is an image that was created by man.

God Wants You Separate From The World!

It is important to realize that God's description of what the earth looks like does not match the description of scientists. God states that the earth is not moving, yet scientists say that it is rotating (Psalms 104:5; Isaiah 45:18). God says, we should not carve graven images of the earth (Exodus 20:4), yet scientists have been carving images of the earth as a ball, (globe) since the Greeks first proposed the ideal (idol). Any ideal that does not agree with God is an idol.

idolatry - image-worship or divine honor paid to any created object. Paul describes the origin of idolatry in Romans 1:21-25 : men forsook God, and sank into ignorance and moral corruption (1:28).

The forms of idolatry are,

- Fetishism, or the worship of trees, rivers, hills, stones, etc.
- Nature worship, the worship of the sun, moon, and stars, as the supposed powers of nature.
- Hero worship, the worship of deceased ancestors, or of heroes

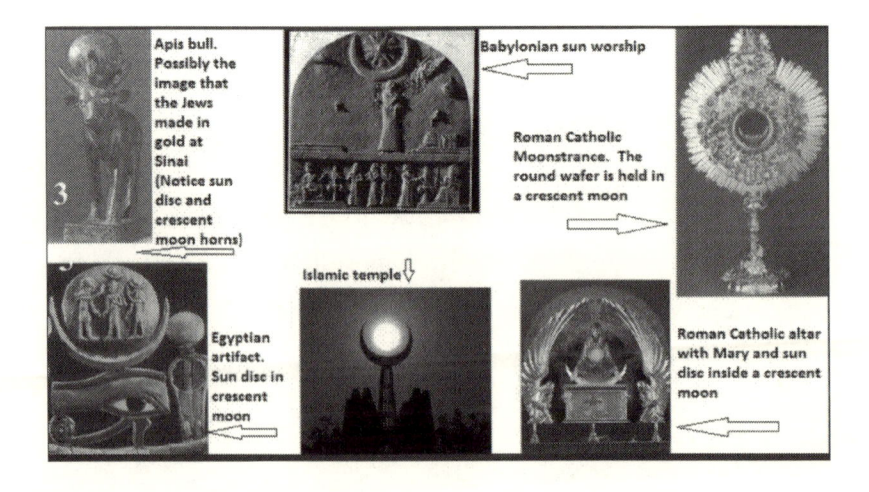

Graven Image

The first and second commandments are directed against idolatry of every form. Individuals and communities were equally amenable to the rigorous code. The individual offender was devoted to destruction (Exodus 22:20). His nearest relatives were not only bound to denounce him and deliver him up to punishment (Deuteronomy 13:20-10), but their hands were to strike the first blow when, on the evidence of two witnesses at least, he was stoned (Deuteronomy 17:2-7). To attempt to seduce others to false worship was a crime of equal enormity (13:6-10). An idolatrous nation shared the same fate. No facts are more strongly declared in the Old Testament than that the extermination of the Canaanites was the punishment of their idolatry (Exodus 34:15 Exodus 34:16 ; Deuteronomy 7 ; 12:29-31 ; 20:17), and that the calamities of the Israelites were due to the same cause (Jeremiah 2:17). "A city guilty of idolatry was looked upon as a cancer in the state; it was in rebellion and treated according to the laws of war. Its inhabitants and all their cattle were put to death." Jehovah was the theocratic King of Israel, the civil Head of the commonwealth, and therefore to an Israelite idolatry was a state offence (1 Samuel 15:23), high treason. On taking possession of the land, the Jews were commanded to destroy all traces of every kind of the existing idolatry of the Canaanites (Exodus 23:24 Exodus 23:32; 34:13; Deuteronomy 7:5 Deuteronomy 7:25; 12:1-3).

In the New Testament the term idolatry is used to designate covetousness (Matthew 6:24; Luke 16:13; Colossians 3:5; Ephesians 5:5).

These dictionary topics are from
M.G. Easton M.A., D.D., Illustrated Bible Dictionary,
Third Edition, published by Thomas Nelson, 1897. Public Domain copy freely.

Babylonian gods

The word idolatry is from the word "idea" (or idol). It means to create an image in your mind. Usually these images are ideas that come from the spiritual world. The Eastern Bible dictionary states that, Idolatry is image-worship or divine honor paid to any created object. Paul describes the origin of idolatry in Romans 1:21-25: men forsook God and sank into ignorance and moral corruption. In fact, any idea or belief that doesn't align with the teachings of God is false and indulging in such practice is idolatry. These practices are dangerous, and if not controlled will be exalted above God, Himself. The Bible states that we should cast down ideas and vain imaginations that exalt themselves above the word of God (2 Cor. 10:5-7). Therefore, even simple hypotheses and theories made by scientists must follow the revelation of God; otherwise, they are idols.

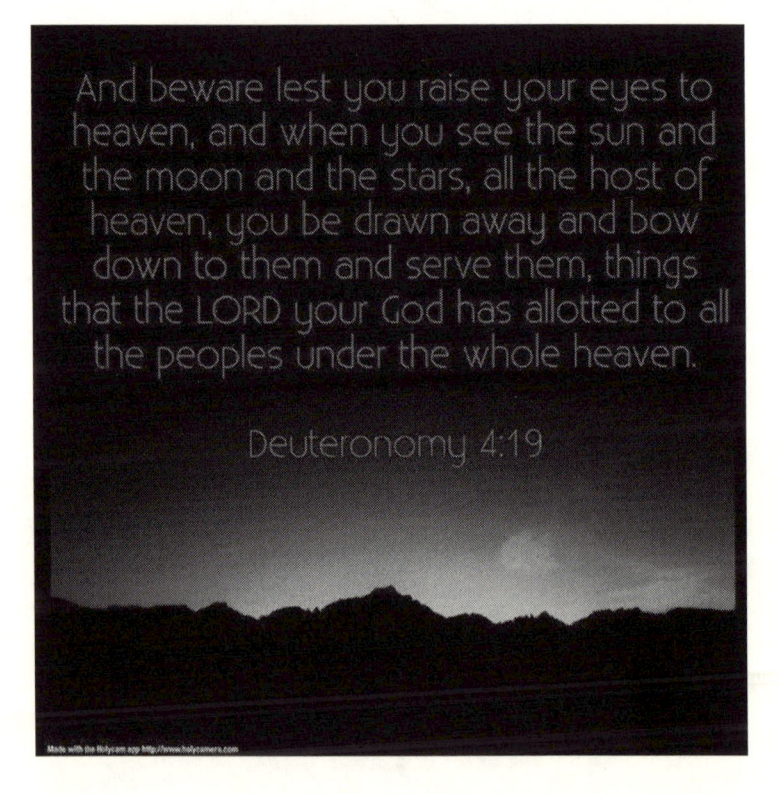

> And beware lest you raise your eyes to heaven, and when you see the sun and the moon and the stars, all the host of heaven, you be drawn away and bow down to them and serve them, things that the LORD your God has allotted to all the peoples under the whole heaven.
>
> Deuteronomy 4:19

This is quite different from the Old Testament view of idolatry. Most people think that idols are only physical in nature, yet this is only partially true. Idols are not merely images made from wood or stone, but idols are ideas in the spiritual world that have access to our minds. This is the reason for me writing this book. God is a jealous God, who requires that we believe everything that he has said in his word. Failure to do so, means that we do not believe God. That' a huge problem. 2 Corinthians 10:5 states that we are to cast down the idols of our mind.

There is a huge difference between God's revelation of our cosmology as indicated by the word of God, and the cosmological view of modern-day scientists. So vast is the difference that it is glaringly obvious that the scientific view of cosmology is in opposition to the biblical view of cosmology.

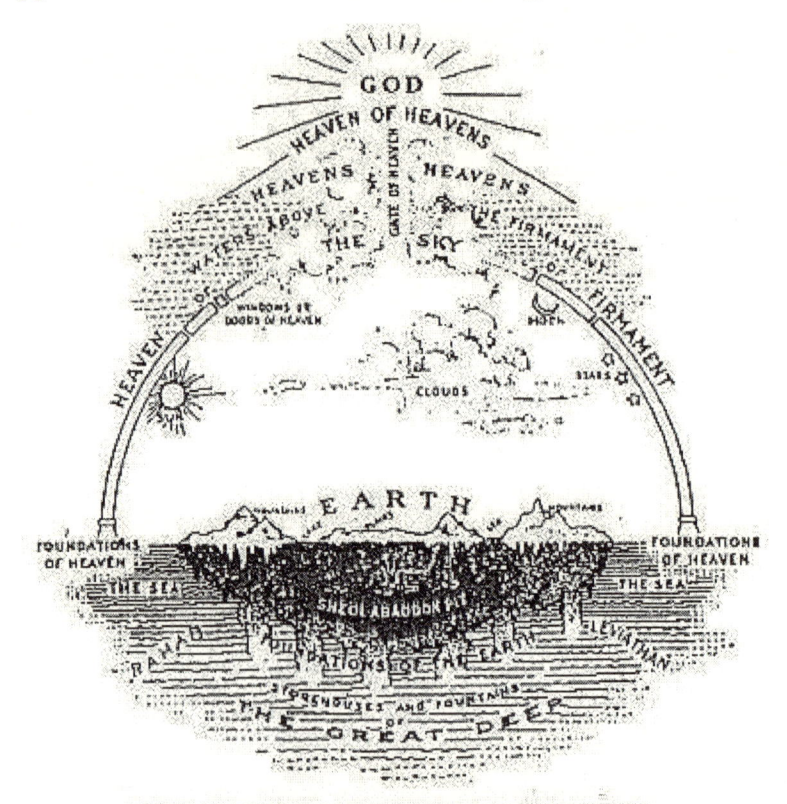

THE ANCIENT HEBREW CONCEPTION OF THE UNIVERSE

TO ILLUSTRATE THE ACCOUNT OF CREATION AND THE FLOOD

Illustration from George L. Robinson, *Leaders of Israel* (New York: NY: Association Press, 1913), p. 2.

Reprinted with permission from A.J. Mattill: The Seven Mighty Blows to Traditional Beliefs, by A.J. Mattill.: Flatwoods Free Press, 1995."

Paul warned his young pastor and protege, Timothy to all oppositions of science in preference to the revealed word of God. Notice what Paul says, O Timothy, keep that which is committed to thy trust, avoiding profane vain babblings, and oppositions of science falsely so called: (1Tim. 6:20).

The 2000 version of the Jubilee Bible states it like this, O Timothy, keep that which is committed to thy trust, turn away from profane voices and vain things and arguments in the vain name of science... In other words, Timothy was told to not allow science to change his understanding of the truth as revealed to him by God. The second thing to consider here; however, is that most science is pagan and almost always based on the thoughts of men, rather than the divine revelation of God.

With this in mind, we approach the topic of why the earth is not a globe. The globe is a spinning ball suspended in the void of space, to which mostly all humanity believe is the true nature of our world. This is something so ingrained in the human psyche, that only a madman would dare to question its reality. Most people would argue that you were an unscientific idiot to question such a truth, indeed! Yet, God's word emphatically reveals the opposite.

According to the word of God, the earth is dry land that has been stretched out above the seas. The earth is not land and sea together.

According to the book of Genesis, "And God said, Let the waters under the heaven be gathered together unto one place, and let the dry land appear: and it was so."

GENESIS 1:10

KING JAMES VERSION (KJV) — FACEBOOK.COM/TOHHBIBLEVERSES

AND GOD CALLED THE DRY LAND EARTH; AND THE GATHERING TOGETHER OF THE WATERS CALLED HE SEAS: AND GOD SAW THAT IT WAS GOOD.

Most people believe that science and the Bible are referring to the same thing when mentioning the earth; but they are not. The word earth means something totally different in the Bible. According to the Bible, the sea is not part of the earth. The sea is one thing and the earth is another. If you recall, the Bible explains that in the beginning the Spirit of God hovered above the waters, but the earth was not formed yet. "And the earth was without form, and void; and darkness was upon the face of the deep. And the Spirit of God moved upon the face of the waters" (Genesis 1:2) KJV. In other words, the sea was here long before the earth was formed. The earth is dry land; and the sea is water. Science says the earth is a globe, with both waters and land joined. This is false.

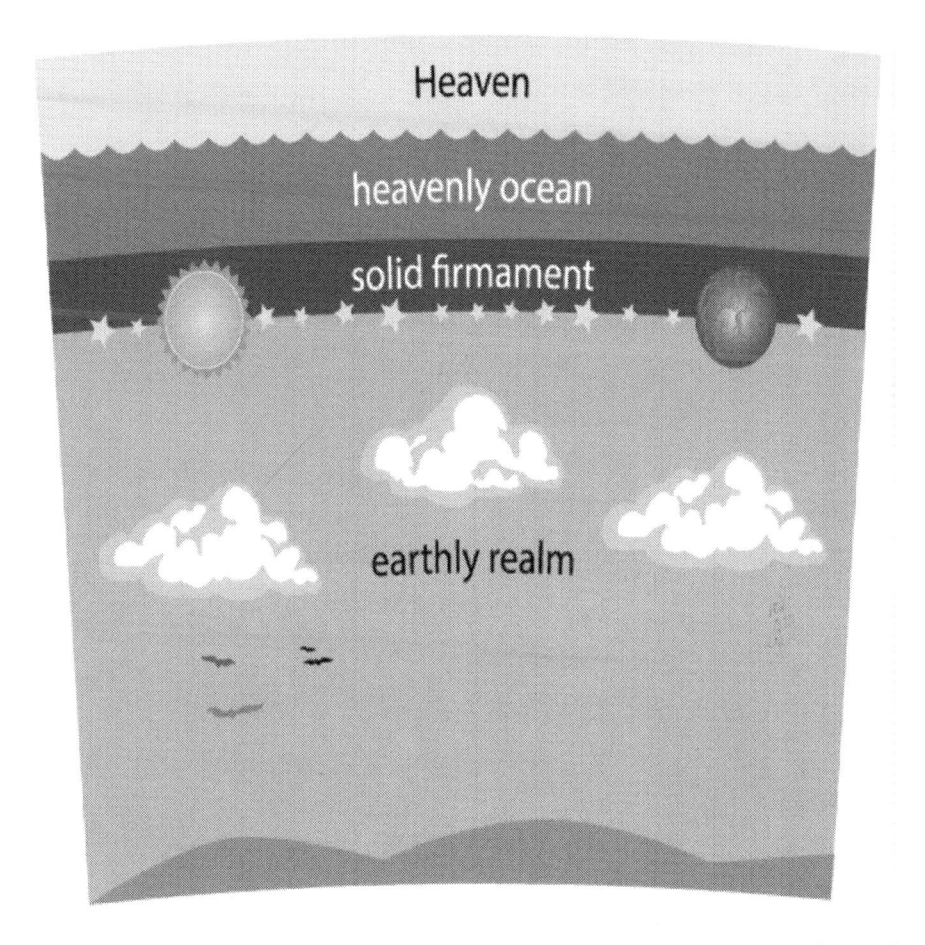

[Diagram: Heaven / heavenly ocean / solid firmament / earthly realm]

And God called the dry land Earth; and the gathering together of the waters called he Seas: and God saw that it was good." (Genesis 1:9-10) KJV. Furthermore, the word of God reveals that the earth rest on pillars, surrounded by the waters; so, there is no such thing as outer space, gravity, and a spinning ball in the void of endless space. (Genesis 1:9-13, Job 9:6, Psalm 75:3, I Samuel 2:8, I Samuel 14:5). When one thinks about it, it is truly mind blowing that there exists such a massive difference between science and the word of God. Yet, for over 500 years, the minds of humanity have been commandeered into accepting a philosophical viewpoint of our cosmology that is in direct opposition to the revelation of our cosmology as revealed in God's word. Now, I realize that this is a difficult topic to approach, since many people will assume that the Bible is not a book of science, and somehow there is a way to think secular and still serve God. Yet, for the few of you who know that God really meant what he said when he said, ...Thou shalt love the Lord thy God with all thy heart, and with all thy soul, and with all thy mind (Matt. 22:37) KJV, you are the ones who this book addresses. Please pay attend to the last words of our Lord Jesus. You must serve God with all of your mind. This means that you cannot serve God in your private devotions, while leaving your mind under the influence of science that opposes God's word. This book is for those of you who understand that you cannot serve God without your mind.

Therefore, Satan's chief goal is to control your mind.

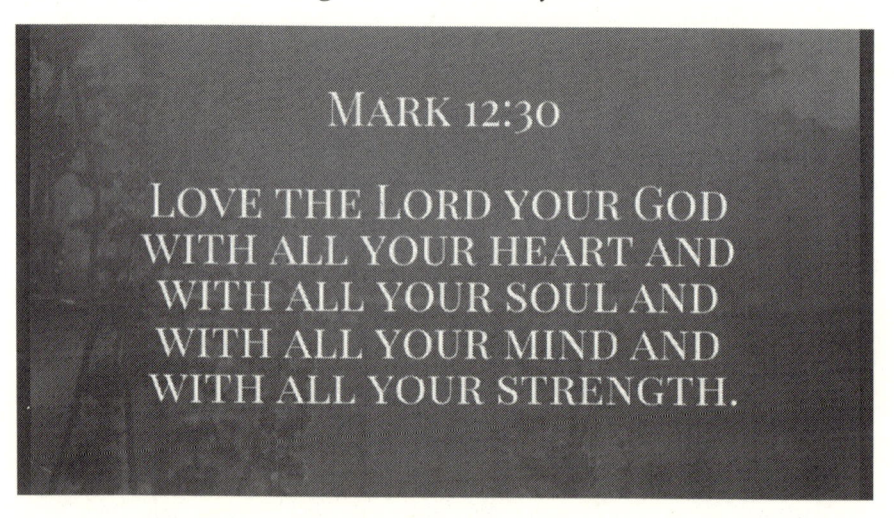

> **MARK 12:30**
>
> **LOVE THE LORD YOUR GOD WITH ALL YOUR HEART AND WITH ALL YOUR SOUL AND WITH ALL YOUR MIND AND WITH ALL YOUR STRENGTH.**

> Serving the LORD with all humility of mind, and with many tears, and temptations, which befell me by the lying in wait of the Jews:
> —Acts 20:19

Throughout reading this book, I will post questions at the end of each chapter designed to help you with this very difficult topic. Please take the time to look up the scriptures and prayerfully contemplate their implications.

Group Study:

1. According to the reading, what does the word idolatry mean, and where do ideas originate?
2. What should we do with ideas that don't line up with the word of God according to? (2 Cor. 10:5-7)

3. What was Timothy instructed to do with scientific views that were in opposition to the word of God? (1Tim. 6:20)

4. Draw a scientific depiction of the earth.

5. Read Genesis 1:6-13, Job 9:6, Psalm 75:3, I Samuel 2:8, I Samuel 14:5 and draw a picture of what it describes.

6. What are the physical differences between how God describes our earth and how modern day scientists describe our earth?

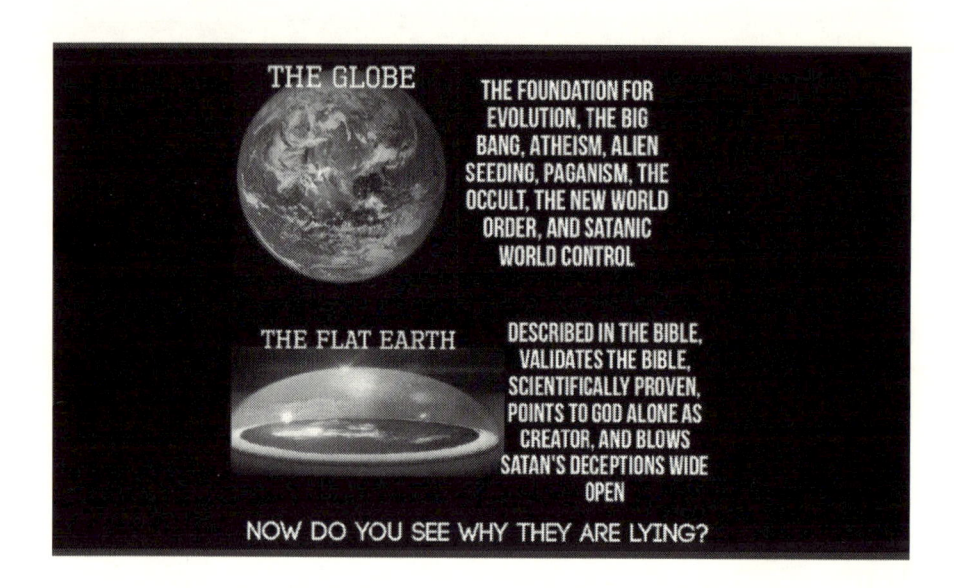

THE GLOBE — THE FOUNDATION FOR EVOLUTION, THE BIG BANG, ATHEISM, ALIEN SEEDING, PAGANISM, THE OCCULT, THE NEW WORLD ORDER, AND SATANIC WORLD CONTROL

THE FLAT EARTH — DESCRIBED IN THE BIBLE, VALIDATES THE BIBLE, SCIENTIFICALLY PROVEN, POINTS TO GOD ALONE AS CREATOR, AND BLOWS SATAN'S DECEPTIONS WIDE OPEN

NOW DO YOU SEE WHY THEY ARE LYING?

CHAPTER 1
The Hebrew View of the Earth

According to the Bible, water covers the dome of the Raqia. In Genesis 1:6; And God said, "Let there be a firmament (Raqia) in the midst of the waters, and let it divide the waters from the waters."

The word Raqia means a solid firmament or a wide expanse. The verse is very clear, God created the Raqia, and he divided the waters which were below the Raqia from the waters above the Raqia, and it was so. He then named the Raqia above, "Sky" (heavens) and there was evening and there was morning on the second day" – Genesis 1:6:8.

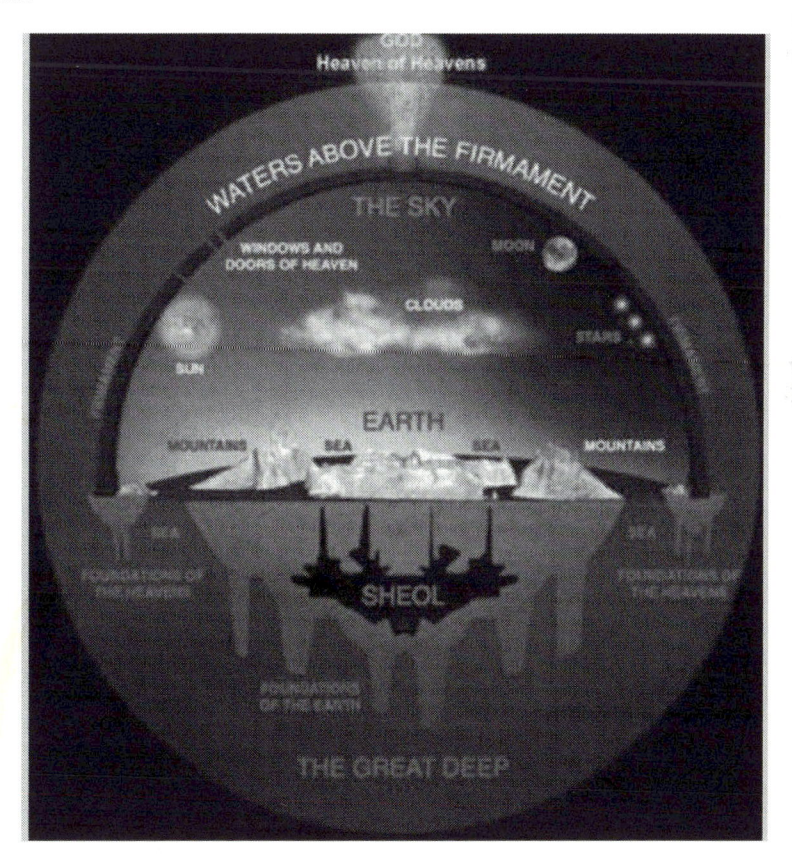

For interpreters of the Bible, this singular verse has been very troublesome; and some questions have been; what does Raqia mean and what exactly were these waters that it divided? Or could this be the atmosphere? However, looking at this passage and the context, one must pause to wonder; what is so special about these events that a whole day was dedicated to it? Furthermore, one must also ask, how does the creation of the firmament build the foundation for the next four days of creation? Obviously, God used the preceding days of creation as a guide for what would come afterwards. Regardless of the modern interpretation of scripture, it is pretty apparent that the Hebrew people of the Old Testament believed the heavens were made of some sort of solid material.

Considering that the origin and extensive use of the English word "firmament" establish the existence of a cosmic "vault" or another body or object similar to a crystalline celestial sphere (maybe or maybe not) as we have believed. The word "Firmament" is translated from a 1600 Vulgate Latin word Firmamentum.

The Vulgate itself was guided by the Septuagint's στερέωμα (stere ō ma), the Septuagint being a 2300- year-old translation. Stere ō ma originates from the word στερεόω (stereo ō) meaning "to make or be firm or solid." Even though secular scholarship and conservative belief try to define the original intent of the Bible and how it was/is interpreted regardless of how old the interpretation is/were. However, it is noteworthy that as early at 250 B.C, the Hebrew Bible word Raqia was translated as a type of solid heavenly entity.

Whether some think this is all a matter of interpretation or not, this concept of the ancient Jewish model of cosmology leaves little room for any other interpretation, even though many scholars from the evangelical community still question its existence. For example, many Christians based their views on the translation of Raqia as "an expanse instead of a firmament or dome. Even the NIV (1984) uses the word expanse: Let there be an expanse between the waters

NRSV: Let there be a dome in the midst of the water.

Those who argue the Raqia as an expanse believe that it represents the atmosphere and not a solid cosmic vault. However, the bottom line of their argument is that Raqia also means solid as well as expansion but prefer Raqia to be translated as an expanse than a firmament or cosmic vault.

Nonetheless, those championing the translation of Raqia as a firmament based their argument of the etymology of the word, and proof from contextual evidence and history.

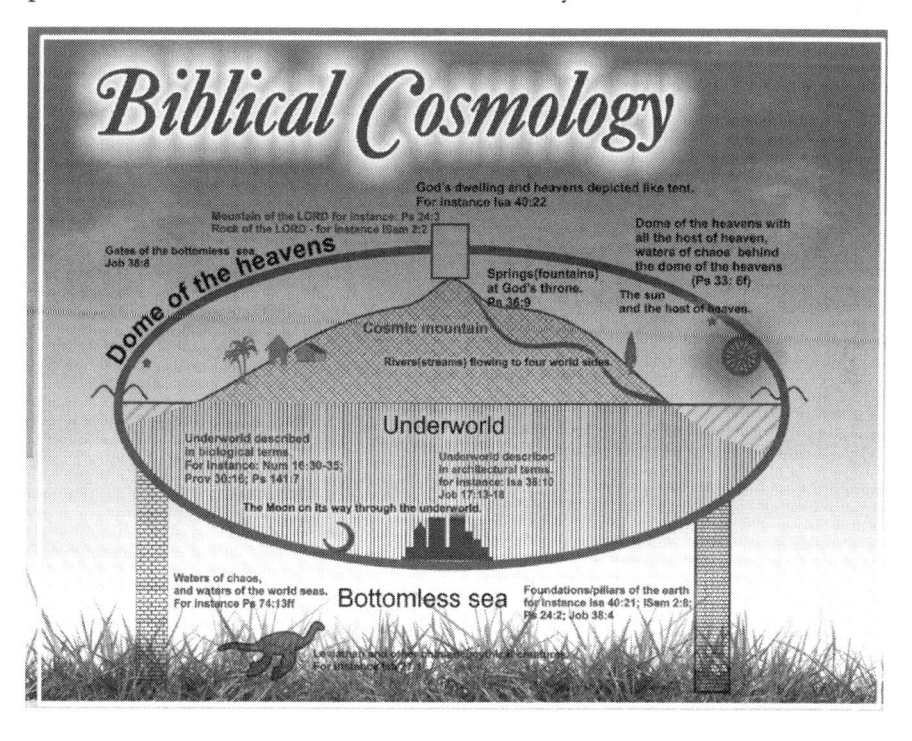

Creation

The creation story begins before anything exists except for God himself. In Genesis 1, the very first chapter of the Bible, we read how the Spirit of God moved over the face of the deep waters. Right away we are forced to envision an earth made of nothing but water. This picture of water everywhere is very difficult to imagine as spherical, because water naturally lies flat.

He created light on the first day, next he created a firmament to separate the waters from the waters on the second day, then he created land on the third day, sun and moon on the fourth day, birds and water animals on the fifth day, animals and man on the 6th day. God then called His creation good and on the seventh day and then God rested.

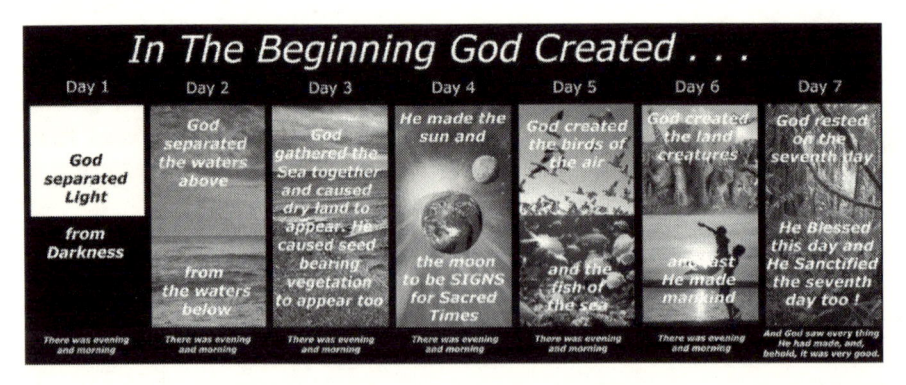

Day 1	Day 2	Day 3	Day 4	Day 5	Day 6	Day 7
God separated Light from Darkness	God separated the waters above from the waters below	God gathered the Sea together and caused dry land to appear. He caused seed bearing vegetation to appear too	He made the sun and the moon to be SIGNS for Sacred Times	God created the birds of the air and the fish of the sea	God created the land creatures and last He made mankind	God rested on the seventh day. He Blessed this day and He Sanctified the seventh day too!
There was evening and morning	There was evening and morning	There was evening and morning	There was evening and morning	There was evening and morning	There was evening and morning	And God saw every thing He had made, and, behold, it was very good.

To understand the biblical record of the creation, one must do away with two red herrings of science. One is the theory of evolution and the other is the theory of gravity. Never once did God say that he took millions of years to create the earth as evolutionist teach, nor did He say that He made the earth spherical, and used gravity to keep people from falling off as taught in the Newtonian model of the earth. Instead, God commanded the dry land to appear and called the dry land earth. "In the Newtonian astronomy, continents, oceans, seas and islands are considered as together forming one vast globe of

25.000 miles in circumference. This assertion will be seen to be entirely false, contrary to the plain, literal and manifest teachings of the Word of God" (Alex Gleason, 1893).

According to the Bible, the Earth is dry land and not seas and lands together as pictured by Sir Isaac Newton. The biblical record states this, "And God said, 'Let the waters under the heaven be gathered together unto one place, and let the dry land appear.' And God called the dry land Earth, and the gathering together of the waters called He seas." (Gen. 1: 9-10).

One would think that if the Earth were a globe, God would not say, let the waters be gathered together in one place and let the dry land appear, because scientists teach that the earth was a rock without water first and that it took millions of years to develop water on the earth. In God's creation account, God reveals that water was here first.

Also, if the earth were a globe, God could not tell the waters to be gathered together, since, according to the globe model, the waters are more boundless than land. The average globe depicts the land gathered together, but the waters abounding everywhere. Finally, if the earth were a globe, God would not be silent about gravity, since gravity would be so essential that God would at least mention it at least once in the biblical record.

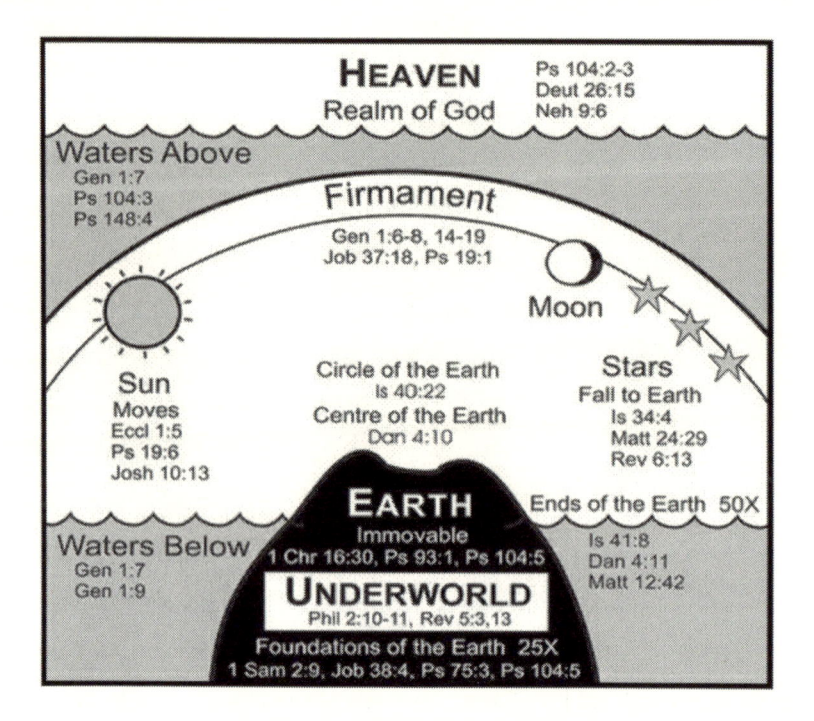

```
              HEAVEN        Ps 104:2-3
                            Deut 26:15
            Realm of God    Neh 9:6

Waters Above
Gen 1:7
Ps 104:3        Firmament
Ps 148:4
                Gen 1:6-8, 14-19
                Job 37:18, Ps 19:1

                                  Moon

           Circle of the Earth   Stars
  Sun        Is 40:22          Fall to Earth
 Moves      Centre of the Earth  Is 34:4
 Eccl 1:5    Dan 4:10           Matt 24:29
 Ps 19:6                         Rev 6:13
 Josh 10:13
              EARTH          Ends of the Earth  50X
              Immovable
Waters Below  1 Chr 16:30, Ps 93:1, Ps 104:5  Is 41:8
Gen 1:7                                        Dan 4:11
Gen 1:9     UNDERWORLD                         Matt 12:42
            Phil 2:10-11, Rev 5:3,13
          Foundations of the Earth  25X
          1 Sam 2:9, Job 38:4, Ps 75:3, Ps 104:5
```

II. The Structure of Our World

The Hebrew cosmology also describes a three tier system, which is a reflection of the fundamental components in the beliefs of all Mesopotamia philosophies of old. In this idea, heaven, earth and Hades, are described as a three-tier structure. Heaven is a solid vault (Ps 19:2[1], which holds the raging waters above from overwhelming the earth (Gen 1:6-8; Ps 148:4). To the firmament, the lights, the sun, moon, and stars were added (Gen 1:14-17). However, it has openings where the waters in the firmament can pass through and flood the world (Gen 7:11; 8:2; 2kings 7:2,[19]. In verses 2 Sam 22:8 and Job 26:11, the firmament, a big bell-shaped is held in place by solid foundation (pillars, analogous to i š i d š a m e of the Babylonians) just like the earth (Ps 75"4[3]; 104:5; Job 9:6 _) and the mountains (Ps 18:8[7]. (G. Bartelmu s, שָׁמַיִם " TDOT 15:211)

III. The Firmament of the Earth

Heaven is used to describe firmament in the Hebrew Bible (Heb Raqia) as a dome-shaped object that covers the earth separating the waters above from the water below..... While the firmament can also be a casing over the earth and heaven including the firmaments itself. (Mitchell G. Reddish, "Heaven," Anchor Bible Dictionary 3:90.)

Following the consensus view of the firmament concept, I will use the word firmament when speaking on that idea and on the other hand, I will use the word Raqia and the alias Raqia when referring to the scriptural body in a neutral interpretation. However, I must state categorically, that firmament concept is not limited by the number of times the word Raqia was used as it appears 6 times in the bible, where the cosmic vault was used instead.

Subtle Hebrew words šāmayim or its synonyms will replace Raqia. To sum up the evidence for the firmament being a hard dome structure is overwhelming convincing to say the least. Moreover, one might argue that it is not the proper application of biblical exegesis that stands in the way of understanding the reality of the firmament, but the globe, since most theologians simply cannot abandon the scientific view of the cosmos over the biblical one. When it comes down to it, most theologians and religious leaders are afraid to take a stand for God's word.

IV. The Pillars of the Earth

The Bible also talks about the pillars of the earth. In Job 9:6 it says, "Who shakes the earth out of its place, and its pillars (ydwmu) tremble." The LXX says, "Who shakes the earth under heaven from its foundations and its pillars (stuloi) totter." In Psalm 75:3 it says, "The earth and all its inhabitants are melting away; I set firm its pillars (ydwmu)." The LXX says, "I have strengthened its pillars (stuloi)." In I Samuel 2:8 it says, "For the pillars of the earth are the Lord's and he had set the world upon them." The Hebrew word for pillar is yqxm. The root meaning "to melt" (BDB 1980, 848). Therefore, yqxm means, "a molten like pillar." The only other place it occurs is in I Samuel 14:5 referring to a mountain. Probably the pillars of the earth are the same thing as the foundations of the earth which were mountains. (Institute for BIblical and Scientific Studies, updated 2019)

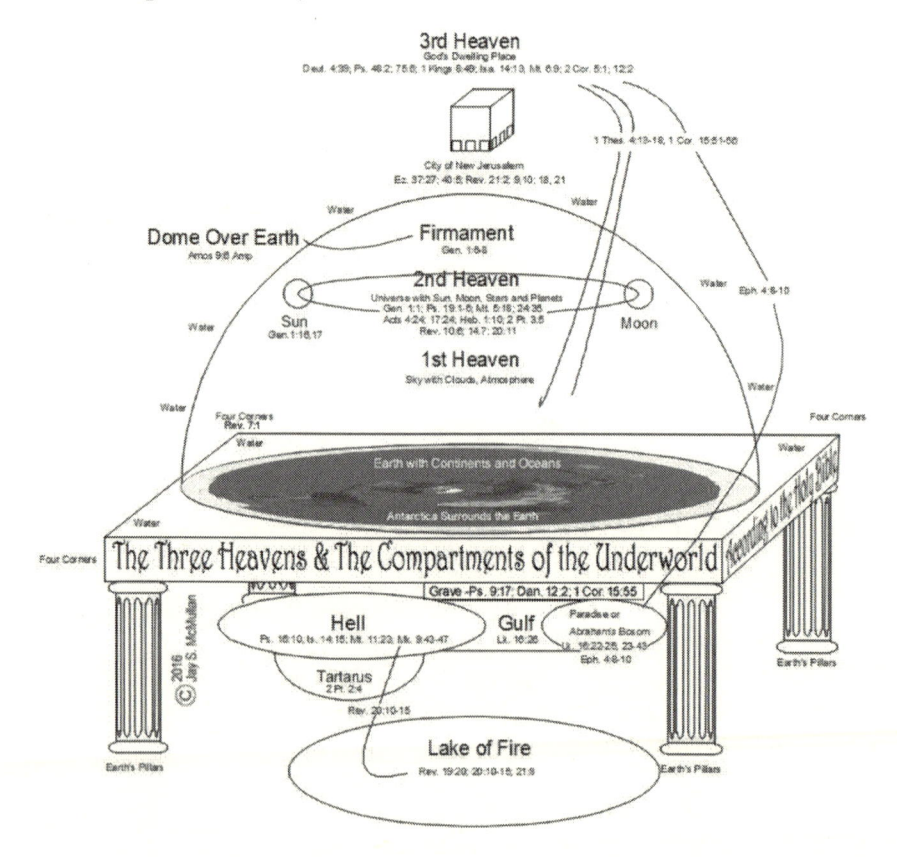

[Figure: Diagram showing a flat earth with a dome above, pillars of the earth supporting it, HELL between the pillars, and "The Foundations of the earth" beneath.]

V. Extra Biblical Beliefs about the Structure of our Earth.

Many other cultures worldwide record our world is a three tier structure, encompassing, a heavenly realm, earthly realm, and an underworld. Recent discoveries such as the Dead Sea scrolls and the Ugaritic text have revived our curiosity about the true nature of our world. For example, in Ugaritic we have seen that there are two mountains, trgzz and trmg that bind the earth. Gibson says that these twin mountains were founded in the earth-encircling ocean, and held up the firmament, and also marked the entrance to the underworld (1978, 66). The mountains are said to bind the earth.

This may indicate that the mountains surrounded, and supported the earth as well as confine the netherworld. The mountains were seen as the foundations of the earth, and the support pillars for the heavens.

The Hebrews held a very similar view as the verses above indicate, as well as later Hebrew writings. So, the phrase "pillars of heaven" and "pillars of earth" are referring to the same mountains. One emphasizes the height of the mountains holding up heaven, the other emphasizes the depth of the mountains that hold the earth firm (Institute for Biblical and Scientific Studies, updated 2019).

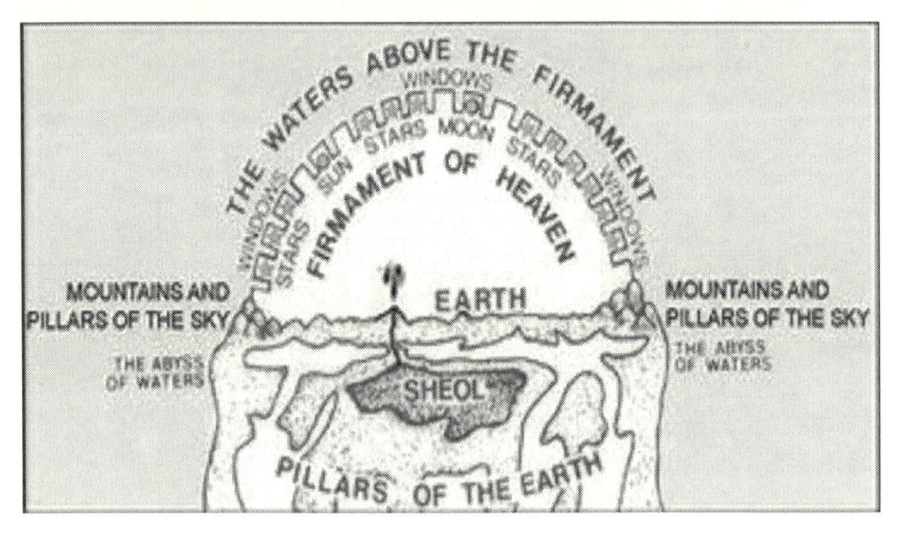

It is clear that the ancient Hebrew cosmology is significantly different than our current view of the earth. The scientific model has all but eradicated the biblical view of the earth. The globe is the problem.

When people imagine the globe spinning in space against the dark backdrop of an endless void, it becomes impossible to reconcile the Biblical model of a three tier flat earth stationary system that sits on pillars, with a scientific one. Since there is no room for both models in your belief system, you are forced to choose which of these two models of the earth you will believe. While this might seem like an issue of scholarship, it actually is a matter of faith. Do you have the faith to believe that God's word is true despite what science says?

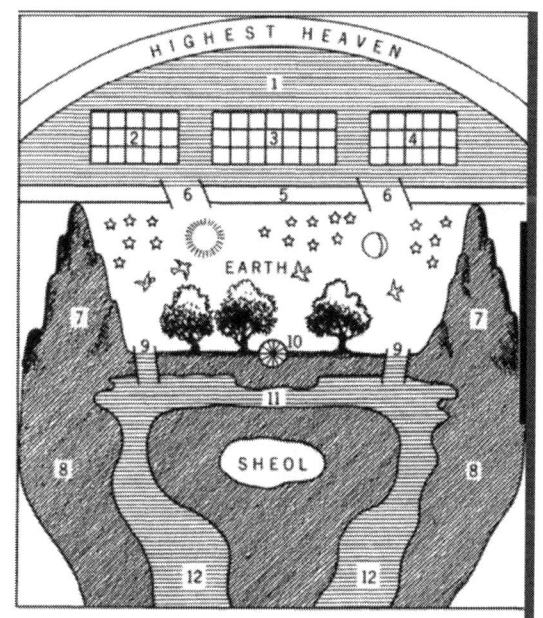

Biblical conception of the world: (1) waters above the firmament; (2) storehouses of snows; (3) storehouses for hail; (4) chambers of winds; (5) firmament; (6) sluice; (7) pillars of the sky; (8) pillars of the earth; (9) fountain of the deep; (10) navel of the earth; (11) waters under the earth; (12) rivers of the nether world

Group Study

1. What is the Earth according to Genesis 1: 9-10?

2. According to the Genesis 1: 9-10, can the earth be part sea and land? (Why or Why not)?

3. What does the word firmament (Raqia) literally mean?

4. What was the general purpose of the (Raqia)? (Genesis 1:7).

5. List a few other scriptures that describe the firmament, besides Genesis 1:6

6. According to Exodus 20:4, interpretation of cosmology, what are the three tier structure of our world?

7. Where are the sun, moon, and stars according to the Hebraic scriptures?

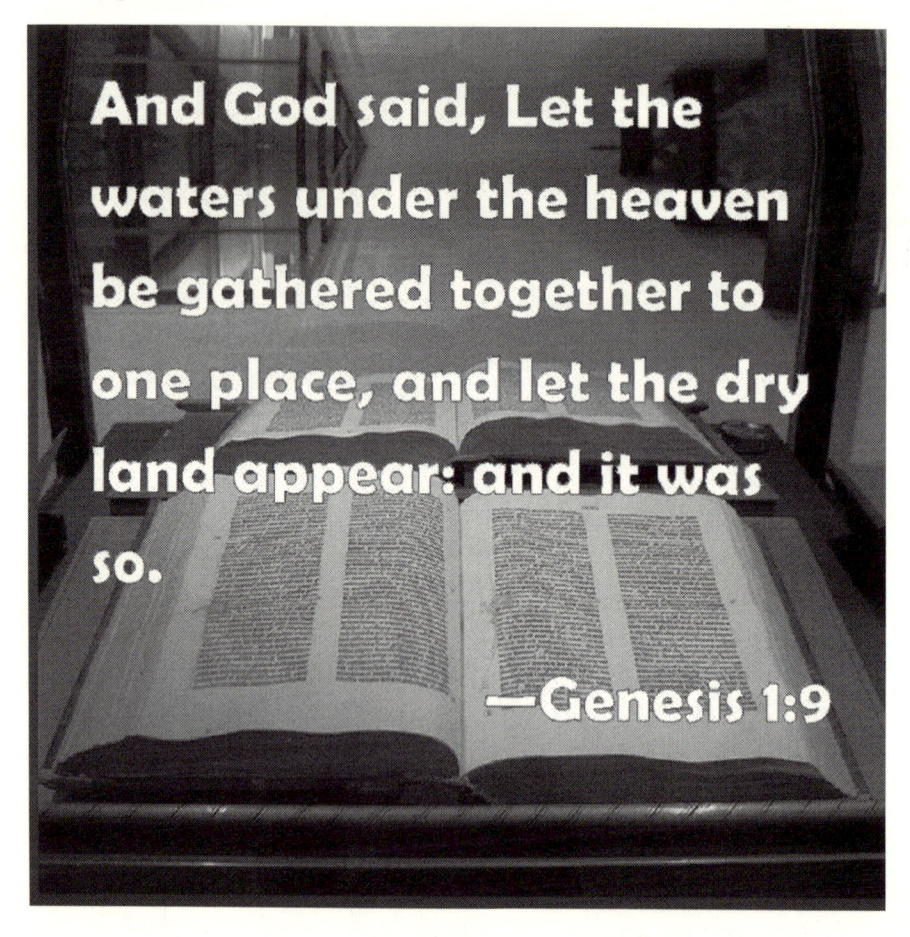

> And God said, Let the waters under the heaven be gathered together to one place, and let the dry land appear: and it was so.
> —Genesis 1:9

Ancient Hebrew Conception of the Universe

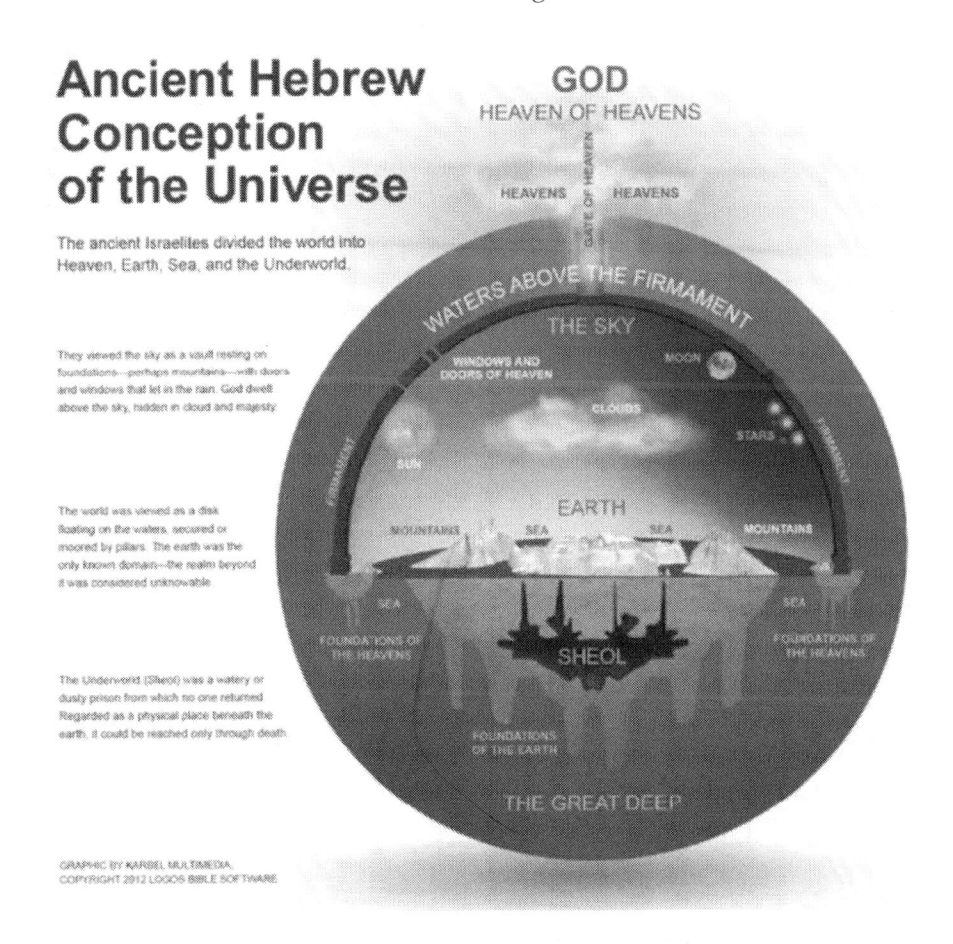

The ancient Israelites divided the world into Heaven, Earth, Sea, and the Underworld.

They viewed the sky as a vault resting on foundations—perhaps mountains—with doors and windows that let in the rain. God dwelt above the sky, hidden in cloud and majesty.

The world was viewed as a disk floating on the waters, secured or moored by pillars. The earth was the only known domain—the realm beyond it was considered unknowable.

The Underworld (Sheol) was a watery or dusty prison from which no one returned. Regarded as a physical place beneath the earth, it could be reached only through death.

GRAPHIC BY KARBEL MULTIMEDIA.
COPYRIGHT 2012 LOGOS BIBLE SOFTWARE

Important Scriptures to Remember From GeocentricWorks.Com

The Four Corners of the Earth:

Isaiah 11:12 "..... four corners of the earth..."

Revelation 7:1 "..... four corners of the earth..." Revelation 20:8 "..... four corners of the earth..."

Still Earth

If the Earth isn't moving then earth cannot be spinning or orbiting around the sun 1 Chronicles 16:30 ".....the world also shall be stable, that it be not moved..."

Psalm 96:10 " ...the world also shall be established that it shall not be moved..." Psalm 93:1 " ... the world also is stablished, that it cannot be moved..."

Pillars

Samuel 2:8 "..for the pillars of the Earth are the Lord's, and he hath set the world upon them...." Job 9:6 "...and the pillars thereof tremble..."

Psalm 75:3 "...I bear up the pillars of it ..."

Firmament Dome

Isaiah 24:18 "...For the windows of heaven are opened..." it's a dome over a flat earth

Isaiah 13:13 "..Therefore I will make the heaven tremble..." if the heavens can tremble it must be solid Genesis 1:6 "....and God said, Let there be a firmament in the midst of the waters, and let it divide the waters from the waters." This is why all the stars rotate together. The stars were placed in the firmament dome

Isaiah 44:24 "... I am the Lord, who made all things, who alone stretched out the heavens, who spread out the earth by myself..."

Psalms 18:9 "...he bowed the heavens also...."

Samuel 22:10 "...he bowed the heavens also..." Strong's H5186 bowed = bend. The heavens bend around us, it's a domed heaven over a flat earth.

The Sun

Job 11:8 "...it is as high as heaven; what canst thou do? deeper than hell; what canst thou know?

Job 11:9 The measure thereof is longer than the earth, and broader than the sea.

Wisdom

Colossians 2:8 "..beware lest any man spoil you through philosophy and vain deceit, after the tradition of men, after the rudiments of the world..."

1 Timothy 6:20 "...false science....

1 Corinthians 3:19 "... For the wisdom of this world is foolishness with God..." A ball shaped world is wisdom

Day & Night

Job 26:10 "He hath compassed the waters with bounds, until the day and night come to an end."

Foundations

Job 38:4 "...where wast thou when I laid the foundations of the earth? declare, if thou hast understanding..."

As we have learned the foundation of earth are the pillars.

Isaiah 48:13 "...mine hand also hath laid the foundation of the earth, and my right hand hath spanned the heavens: when I call unto them, they stand up together..."

Footstool

Isaiah 66:1 "the heaven is my throne and the earth is my footstool..." Matthew 5:35 "...nor by the earth; for it is his footstool..."

Acts 7:49 "...heaven is my throne, and earth is my footstool..."

Desktop Globe

Exodus 20:4 "...thou shalt not make unto thee any graven image, or any likeness of any thing that is in heaven above, or that is in the earth beneath..."

Isaiah 42:8 "...neither my praise to graven images..." This is the 2nd commandment Stars

Matthew 24:29 "...the stars shall fall from heaven..." Mark 13:25 "....and the stars of heaven shall fall..."

Revelation 6:13 "And the stars of heaven fell unto the earth..." earth.

Universe

The ancient Hebrew view of the universe looks like this:

The earth is on pillars, the sheol under the land, a dome over the flat geostationary earth with all the stars, sun and moon inside the dome going around the earth. That's why in Genesis 1:14 God said, "... let there be light in the firmament of heaven". Not orbiting a Milky Way galaxy.

Sun Circuit

Psalm 96:10 "...the world also shall be established that it shall not be moved."

Psalm 19:6 "His going forth is from the end of the heaven, and his circuit unto the ends of it and there is nothing there hid from the heat thereof."

Circle of the Earth

Isaiah 40:22 "It is he that sitteth upon the circle of the earth"

Isaiah knew the difference between a circle and a sphere

Isaiah 22:18 "...toss thee like a ball..." So why didn't Isaiah use the same term for the ball earth that he did to describe a ball? That is because the earth is not a ball. The earth is a flat disc.

Grasshoppers

Psalm 33:14 "...from the place of his habitation he looketh upon all the inhabitants of the earth..." You can't do that on a 13.8 billion lightyear across the universe. The earth has a dome.

Isaiah 40:22 "...it is HE who sits above the circle of the earth, And its inhabitants are like grasshoppers..."

Deception

Matthew 24:24 "...if it were possible they shall deceive the very elect..." 2 Thessalonians 2:3 "...let no man deceive you by any means..." Ephesians 5:6 "...let no man deceive you with vain words..."

So if you are deceive about the ball shaped earth and believe its the truth then you have no chance at ever being able to understand the bible. You are self barred from all knowledge. If you take the view that the flat earth is a distraction then what exactly is it distracting you from ? Or if you believe it doesn't matter if the earth is a sphere or flat then you are a self-confessed idiot because it absolutely does matter.

Imagine the world where everyone knew there were being watched by the Father. How their behaviors would change knowing they will be held accountable.

Tent

Isaiah 40:22 "It is he that sitteth upon the circle of the earth, and the inhabitants thereof are as grasshoppers; that stretcheth out the heaves as a curtain, and spreadeth them out as a tent to dwell in: How exactly can you confuse a tent, a dome like structure, with a ball ? It's a flat earth with a dome like structure above.

Ends of the Earth

Job 37:3 "...and his lightning unto the ends of the earth."

Daniel 4:11 "...and the sight thereof to the end of all the earth..."

Proverbs 30:4 "..who hath established all the ends of the earth? What is his name, and what is his son's name, if thou canst tell?" WOW - Proverbs is old testament and what a prophecy that is! Not only is it predicting Jesus, the son of God, but it is also describing the ends of the earth. There is no end to a round shaped ball; it goes on continuously. You can only come to the ends of the earth on a flat level plane.

Lies

Jeremiah 16:19 "...the Gentiles shall come unto thee from the ends of the earth, and shall say, surely our fathers have inherited lies, vanity and things wherein there is no profit..." This is a future prophecy following a monstrous destruction of people who believe lies. This could be applied today. Here we are showing evidence of the lies today proving the earth is flat and yet you continue to believe the lies because you inherited the lies. There is no end to the earth on a ball. The earth is flat.

Four Winds

The bible is consistent all the way through from the old testament to the new testament that there are four winds.

Jeremiah 49:36 "...4 winds"

Daniel 7:2 "...4 winds"

Matthew 24:31 "...4 winds"

Revelation 7:1 "...4 winds"

There is a Jetstream that connects North America with Europe. Commercial aircraft try to fly in the Jetstream to conserve fuel.. There are 4 jet streams. Two in the north and two in the south hemisphere. So how did the guys who penned the bible know about these four jet streams? They were not discovered until the 1920's when airplanes could fly high enough to discover them. So we have a dome shaped earth that requires cross ventilation to circulate the air through. If you want to properly ventilate your roof and prevent dry rot or wet rot you need a minimum of two vents. One to intake air and one to vent. If you want to ventilate the entire earth you will need inlets of air and outlets. Revelation 7:1 "...I saw four angels standing at the four corners of the earth holding back the four winds of the earth so that no wind would blow on the earth, the sea or on any tree." Each of them would feed in a Jetstream of air into the dome to cross ventilate the entire earth. Jet streams would not work on a ball shaped earth because they would fizzle out very quickly.

Mathematics

Psalm 147:5 "...his understanding is infinite..."

Genesis 1:1 "...In the beginning God created the heaven (singular) and the earth..." In Hebrew the first sentence of the bible looks like this:

If you take the number of Hebrew letters and multiply that by the product of the letters, then divide that by the number of words times the product of the words you get 3.1416 that's the value of Pie to four decimal places. Pie is the relationship of its "circle" circumference to its diameter, so you measure the circumference of " ANY" circle and divide it by the diameter and you get 3.141 (PIE). What the bible is describing here is clearly a circle and not a sphere. Only a flat geostationary circular earth fits that description.

If that failed to knock your socks off then you haven't been paying attention. What we have got here is the bible describing the creation of earth in words and behind those words we have got a mathematical formula that not only verifies the authenticity of the text (Gods' fingerprint if you will) but it describes the very shape of the earth. Circular 3.141

Luke 18:17 "For nothing is hid, that shall not be made manifest; nor secret, that shall not be known...

**Prove ALL things, hold fast to that which is good.
1 Thessalonians 5:21**

GeocentricWorks.com

CHAPTER 2
The Greeks View of the Earth

It is apparent that almost all our current cosmological understanding of the earth comes from the Greeks. The Greeks developed their understanding to the cosmos from the accumulated knowledge that they gained from the Egyptians and Babylonians. While the Greek understanding of the cosmos differed in many ways, the Egyptians and Babylonians used astronomy for more than just mere grist for debate. For example, the Egyptians relied on well-established calendars to anticipate the flooding of the Nile; rituals were required to be able to tell time during the night, and the orientation of monuments in the cardinal directions was also important. Babylonians believed in the reading of omens in the sky to secure the state. These were all important stimuli to develop their astronomy Ancient History Encyclopedia, 2019). So, in one sense the Greeks followed the Egyptians and Babylonian cosmology, but in another, they too were responsible for adding to the esoteric understanding of today's worldview of the globe.

Greek God Atlas Statue with Globe

They were famous for their schools of higher knowledge, which were rather different than ours. Students would gather around a teacher, perhaps in debate halls, and ask questions and discuss among themselves what might be the answers and the best ways to figure out those answers. They would sit around in open spaces and argue their point of view.

Often students would arrive at their own conclusions through rigorous debate. Even though this seemed like the best way that they could gain knowledge, it often led to confusion and wrong conclusions. This is not surprising, because it was knowledge gained out of pride and the vanity of intellectual prowess. This is why the word of God admonishes humanity not to be proud. God promises that pride always leads us away from God's revelation and truth. Look at, 1 Corinthians 8:1 "...Knowledge puffeth up, but charity edifieth. Also, the writer of Romans states, "

Do not think of yourself more highly than you ought, but think of yourself with sober judgment, according to the measure of faith God has given you" (Romans 12:3).

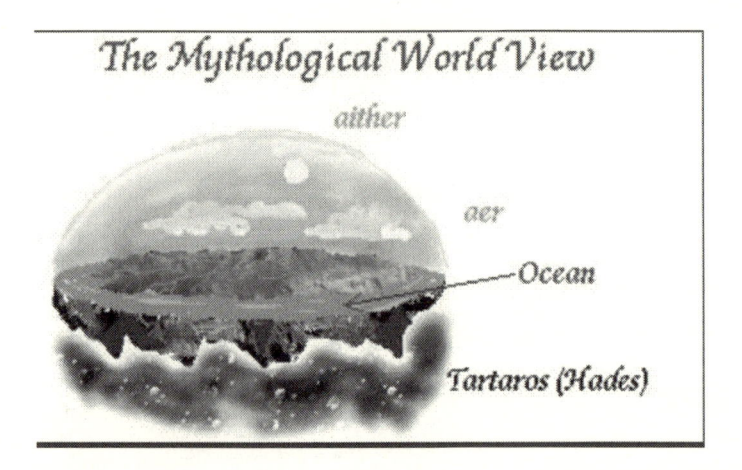

I. Pythagoras (570 – c. 495 BC) - The Occult Origins of the Globe

The occult origins of a globed shaped, non-stationary Earth take us back to a Greek philosopher known as Pythagoras (570 – c. 495 BC). Although many sources credit Philolaus as the one who came up with the model of the Earth, Sun, Moon all rotating around "central fire" in the universe, others claim that Pythagoras held to this idea even before his student Philolaus.

Either way, what is absolutely undeniable is the extremely pagan footing on which both these men stood.

Neither of these men were subscribers to the Hebraic understanding of cosmology; therefore, their view of our cosmos would and can only lead to idolatry, since it replaces Yahweh's word with the vanity of human speculation. To most of us average folks, Pythagoras is presented in school as the ancient Greek mathematician who gave us basic proofs such as the famous "Pythagorean theorem" regarding triangles (A squared plus B squared equals C squared).

To occult adepts however, Pythagoras is understood to be one of the most revered of ancient mystics and Mystery School teachers. To the Freemasons, Pythagoras was a sage in the occult. According to Albert Pike (Grand Commander of North American Freemasonry) his science of numbers was based on Kabalistic principles. "Everything is Veiled in Numbers" To them, "Everything is veiled in numbers" The further one looks into the beliefs and pursuits of Pythagoras, the more it becomes plain that he was not an individual preoccupied with rational and scientific analysis of the natural world, (as our secular education system to often portrays him), Pythagoras was a member of a secret cabal of Gnostics who hid their true intentions behind numbers. Thus, Plato and all the famous Greek Philosophers came after him to be trained in ancient occult number magic. This is how the doctrines of demons were passed down through mystery tradition and ancient secret societies

The further one looks into the beliefs and pursuits of Pythagoras, the more it becomes plain that he was not simply an individual preoccupied with rational and scientific analysis of the natural world, (as our secular education system to often portrays him), Pythagoras was a member of a secret Kabbalah of Gnostics. Indeed, Pythagoras is revered by Plato and all the famous Greek Philosophers who came after him, and in that light we can understand just how pervasively Gnostic all of Greek "Philosophy" truly is, how it was seeded and guided not simply by the imaginations and musings of speculative men, but by the doctrines of demons as they were passed down through Mystery tradition and ancient secret societies.

Pythagoras view of the cosmos became widely accepted that the earth was a sphere in a universe which was itself also fully spherical. This claim was based on pure speculation and the belief that the circle or sphere was the most perfect of geometric shapes, and therefore appropriate for the shape of the earth. One of the so-called proofs for this belief is found in the observation of a ship and its mast. As the vessel receded beyond the horizon, the ship's mast seems to sink over the curve of the earth. While we know this is not true today and merely a matter of perspective, it serves to demonstrate just how unscientific their observations were. Moreover, since it was possible for the ancient Greeks to veil mysteries in numbers through trigonometric considerations, many people simply accepted that Greeks were wise beyond question. This is why Eratosthenes calculations concerning the circumference of the earth was never challenged. Since, at that time, no other assessment of the size and shape of the earth has been offered. It was believed by faith!

II.   Aristotle (384–322 B.C.E.)

Aristotle 384 BC

One such proud Greek who often surmised things of a nature through such foolish rambling was Aristotle. Aristotle was probably who the Holy Spirit referred to through the writings of the Apostle Paul, who wrote, "...Greeks seek after wisdom; but we preach Christ crucified, to the Jews a stumbling block and to the Greeks foolishness, but to those who are called, both Jews and Greeks, Christ the power of God and the wisdom of God" (1 Corinthians 1:22-24).

Six Things That Aristotle Got Wrong

Armand Marie Leroi (2014) writes about six things that the famous Greek philosopher, Aristotle, got wrong. I added one more thing.

1. **Women are monstrous**. Aristotle says that women have fewer teeth than men. Compared to men, he says, they are "immature," "deficient," "deformed"; they are even a bit "monstrous."

2. **Some people deserve to be slaves**.

3. **Eels don't reproduce.** They spontaneously emerge.

4. **The eternity of the world.** This is mere speculation and refutes the Bible.

5. **There's life out there**. Aristotle's believed that the celestial objects are alive.

6. **Bees don't reproduce**, they simply are spontaneously come to life.

7. *The earth is round.*

Since most of their knowledge was speculative and based on ancient Egyptian and Babylonian beliefs, they were at liberty to simply concoct stories of gods and legends as they saw fit. Their speculations are historic and famously repeated in much of today's Greek literature.

Notions about the flat shape of the earth has been around for centuries. Even university professors state that the ancients had many novel ideas about the shape of the earth. The Babylonians thought the earth was hollow, to provide space for their underworld. The Egyptians thought the earth a square, (with four corners) with mountains at the edge supporting the vault of the sky. (Donald E. Simanek, 2006)

However, Aristotle argued for a spherical earth, for these reasons: The gradual disappearance of ships over the horizon, the tops of the sails disappearing last. The shape of the curved shadow of the earth on the moon during eclipses. The variation of the sun's elevation with latitude. (This was the basis of Eratosthenes' measurement.) The variation of a star's elevation with latitude. The fact that one sees new stars as one moves north or south on the earth's surface. Matter tends to form into drops or globs, and the earth, in forming from chaotic matter, did the same (assumptions). Proof by elephants: When one travels west from Greece, one finds elephants (African). When one travels east one finds elephants (Asian). Not realizing that these elephants are different kinds, he thought that one was traveling to the same lands by going in opposite directions. Aristotle made many assumptions, many of which are absolutely absurd. Yet, he remains the leading Greek philosopher of note concerning the globular earth today.

Greeks in the Bible

One famous story in the New Testament records that the Greeks once mistaken Paul and Barnabas for planets (Greek gods). After witnessing the power of God through Jesus Christ to heal, the Greeks at Lycaonian thought Paul and Barnabas must be planets in human form. Barnes Notes on the New Testament tells us: "And they called Barnabas, Jupiter. Jupiter was represented as the most powerful of all the gods of the ancients…" (Acts 14:11). There is the most abundant proof that he was worshipped in the region of

Lycaonia, and throughout Asia Minor. There was, besides, a fable among the inhabitants of Lycaonia that Jupiter and Mercury had once visited that place, and had been received by Philemon. The whole fable is related by Ovid, (Metam. 8, 611). Is it any wonder that the Greeks knowledge of many things, including the Globe was considered foolishness to God?

Listen to what the Lord had to say about the wisdom of men: "For the wisdom of this world is foolishness with God. For it is written, He taketh the wise in their own craftiness" (1 Corinthians 3:19) and "Because the foolishness of God is wiser than men; and the weakness of God is stronger than men" (1 Corinthians 1:25).It is apparent that Paul often had vigorous disputes with men concerning Christ and those considered themselves wise. No wonder Paul warned Timothy, his young pastor in training, not to get tied up with arguments about so called science and the wisdom of his day. Of this, Paul warned.

Timothy: "O Timothy, keep that which is committed to thy trust, avoiding profane and vain babblings, and oppositions of science falsely so called:" (1 Timothy 6:20). Paul stated that it is not even science.

It's a false science that leads to more confusion about the person of Jesus Christ. Paul wanted Timothy to avoid being carried away by this false knowledge from the Greeks and others.

II. The Greeks, Flat Earth, and the Surrounding Oceans

The ancient Greeks developed, over a period of centuries, an elaborate cosmology. The earliest views go back to the time of Homer and Hesiod (the 8th century BC). They postulated that the earth was flat or cylindrical; located in a hemispherical cosmos that surrounded it. Hesiod and Homer, both were flat earth believers, and their cosmology models both were based on a flat disc, surrounded by a world encompassing ocean called "Oceanus". This is expressed no more clearly then by Hesiod himself, in a passage of his lesser-known ascribed work The Shield of Heracles, 314-316: "And round the rim Ocean was flowing, with a full stream as it seemed, and enclosed all the cunning work of the shield. Over it swans were soaring and calling loudly, and many others were swimming upon the surface of the water; and near them were shoals of fish."

Hesiod thus believed the earth was like a flat shield, with an ocean fully surrounding the shield's rim. In his Theogony (700BC), a detailed flat earth cosmography is also present; starting with the creation account from verses 116-138: Hesiod in these passages also noted that Heaven was the "equal" to earth, and this only makes sense on a flat earth model, since if the earth was flat it would be diametric to the above Heaven, therefore it's "equal". Like Hesiod, Homer in his works the Odyssey and Iliad (800BC) also wrote that the earth was the shape of a flat shield or disc. In one verse of the Iliad, Achilles' flat shield is described as having the river Oceanus around its rim, XVIII. 606:

"Therein he set also the great might of the river Oceanus, around the uttermost rim of the strong- wrought shield." It is therefore simply evident that both Hesiod and Homer, the two earliest poets of Greece believed in a flat earth cosmography, specifically that of a disc or circular shield shape, with a world encircling Ocean called "Oceanus". We also find references to this in Strabo (Geo, 1. 1. 3: 7), Aeschylus (Prometheus Bound, 157) and Plato (Phaedo, 112e). (Ancient Belief in Flat Earth, October 10, 2010).

Hesiod's *Theogony* (750 BC)

Cosmology vs. Theogony
- Cosmology: belief that earth was created with purpose by Deity
- Theogony: "birth of the Gods" in Ancient Greek

Hesiod's theory of universal creation from Chaos (cosmic void)
- Gaia - Earth
- Tartarus - underworld (boundary)
- Eros - desire/love
- Erebus - darkness under earth
- Night - darkness above earth

III. Eratosthenes (276 BC-194 BC)

One cannot fully understand that dept of the Greeks influence over our flawed belief that the earth is a globe, until they come to understand just how wrong the assumptions of Eratosthenes were.

Eratosthenes (276 BC-194 BC) was a Greek mathematician, geographer and astronomer.

He was born in Cyrene (now Libya) and died in Ptolemaic Alexandria. He is noted for devising a map system based on latitude and longitude lines and computing the size of the Earth. Konrad Miller, worked on the ancient geographers in Greek, and especially on Eratosthenes. He says that "in the whole antiquity there is only one single measurement of the earth worthy of that name, the one made by Eratosthenes" (published 1919, p.16). According to many writers, all other ancient geographers such as Posidonius, Ptolemy etc. copied Eratosthenes' calculation.

However, one of the most noted Greeks to talk about the size and form of the earth explicitly was Aristotle. He insisted the earth was a globe and reported that mathematicians had given its circumference as 400.000 stades. So essentially Aristotle arrived at this value through hearsay and guessing without knowing anything certain. Then came Eratosthenes. Eratosthenes of Alexandria (276- 195 BC), a Greek mathematician who supposedly determined the size of the earth, through his mathematical calculations without direct measurement.

One Wikipedia article states, "He observed that at noon on the summer solstice of the northern hemisphere, at two different places, the sun's rays fell upon the earth at different angles. In the city of ancient Syrene (modern Aswan, Egypt), at noon on the solstice, the sun was directly overhead and cast no shadows on the objects below it (Aswan is very near the Tropic of Cancer. This was proven by observing that the sun completely illuminated the bottom of a very deep well in Syrene at noon. The angle between the sun's rays and objects perpendicular to the earth's surface was effectively zero

"In Alexandria, on that same day at noon, the sun cast slight shadows behind objects. By measuring the dimensions of the shadows and the objects that cast them, Eratosthenes was able to calculate that the

sun's rays arrived in Alexandria at an angle of a circle (about 0.13 radians or 7.2 degrees). He knew the distance between Alexandria and Syrene to be "stades" from land surveys done between the two cities. A stade was an ancient Greek unit of measurement (150-200 meters). He also assumed that Syrene was due south of Alexandria and on the same meridian of longitude. There are several obvious problems with this:

Eratosthenes measurements were assumptions based on guessing at the unknown distances and measurements, yet we are told in textbooks that he was pretty accurate.

He assumed that the earth was a sphere. He did not know. This was all assumption and speculative at best. Yet, we are told these figures as if they are completely factual.

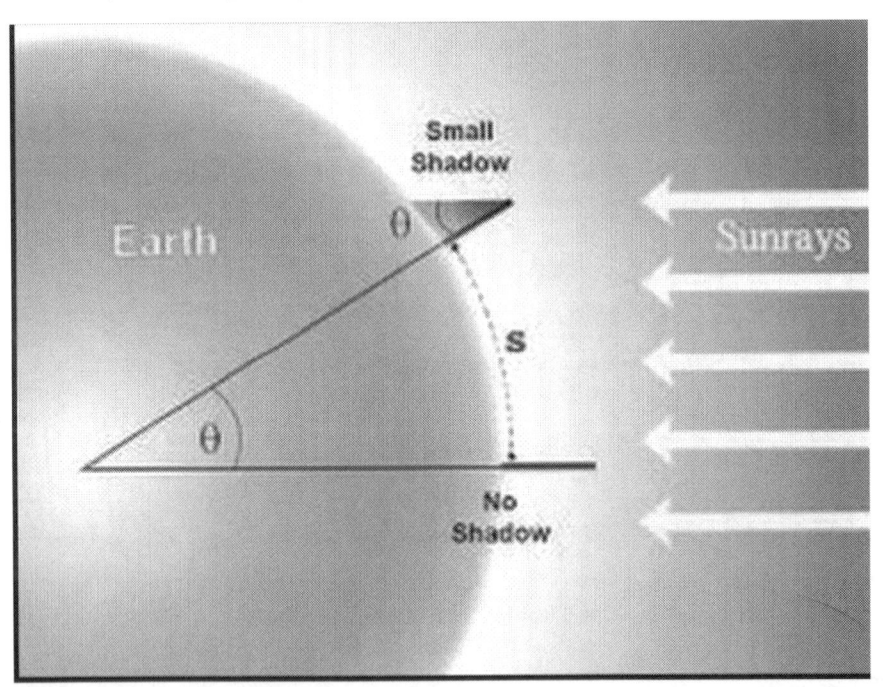

He arrived at the circumference of the earth through the guess work of seamen. "Sailors told him that the distance between Rhodes and Alexandria is about 4000 to 5000 stades. This was their best guess and not an accurate measurement. Any modern observers would know that it is nearly impossible to determine the distance that a ship has sailed by merely guessing.

Eratosthenes neglected the longitudinal difference of 2° and probably used measurements of latitude when he implied a distance of 3750 stades, as Miller says" (p.27). Eratosthenes is given undue credit and glory for this highly inaccurate way of measuring the earth's circumference. Over the centuries of time, Eratosthenes experiment has stood as the primary way that we know, or believe that the earth is 24,901 miles in circumference. Every textbook, discovery channel show, and history class points back to this second century Greek mathematician, just as it does with Aristotle, Pythagoras, and other globalist. The possibility of a second century man guessing the earth's circumference without proper instrumentation to verify such a measurement is impossible.

His books are not preserved, only some contents of the "Book of Dimensions" are quoted in Galen, and other parts mentioned in the "Geographica" of Strabo.

Eratosthenes transformed all of his measures into stades (see also Harley and Woodward, vol.I, p.155). Stades were inaccurate approximations at best. First, a group of royal geodesists measured the distance from Syrene to Meroe in the Sudan (today: Dar Shendy on the Nile), which came to 5000 stades. In this case the longitudinal difference is only 2°, but it is not negligible. And how could they really measure this great distance (about 800 km) over very rough mountainous surface? (Uwe Topper, 2001) Posidonius, who died about 150 years later, chose 4000 stades and arrived at a similarly exact result. Again, this tells me the result was there first, and the way of obtaining it was a pure guess. According to Miller (p.16) recent scholars take this view. They speak of Eratosthenes as "unconsciously" arriving at his results, or borrowing them from another learned culture. (Topper, 2001) The question remains:

Eratosthenes assumed a lot to arrive at his measurements. He assumed the diameter of the earth compared to the sun. He made assumption about the moon as well. Where did Eratosthenes get his these numbers from? According to Eratosthenes, the diameter of the sun is three times that of the earth, its distance is 51 diameters of the earth, and the moon is 19.5 earth-radii away. All figures are far wrong. So if all of his estimates were wrong, how did he arrive at an exact result for the earth's circumference?

The problem of the incorrect data used by Eratosthenes, especially the 3° difference in longitude, is brushed aside by Miller's remarks (p.6 and p.25), that they are corrected by giving the latitudinal difference between Alexandria and Syrene as 7° 1/7 . (Topper, 2001).

Eratosthenes measurements stand firm as the vanguard of a great scientific discovery; even though common sense tells us that his measurements and assumptions were about as accurate as a blind man trying to tell time by looking at the face of his watch. Impossible! So why do modern day scientists continue to push such a foolish story about the shape of the earth? Because they don't know what the earth looks like.

Group Study

1. What was God's view on knowledge according to, 1 Corinthians 8:1?

2. What did the Greeks seek after according to, 1 Corinthians 1:22-24?

3. Why did the Greeks often reject the revelation of Christ? 1 Timothy 6:20

4. What does the term oppositions of science refer to? 1 Timothy 6:20

5. How is the wisdom of this world, foolishness to God? 1 Corinthians 3:19

6. Is God against science? Why or Why not?

7. Can we understand the mysteries God through our own thoughts? Deut. 29:29; 1 Corinthians 2:14.

CHAPTER 3
Men of the Globe

Without the influence of these men, humanity might otherwise had no choice but to accept that the earth is flat. Biblical cosmology overwhelmingly leans to the acceptance of a flat earth, covered by a dome like structure, which we understand to be the firmament. However, it was the influence of the following men, who guided the hearts of men away from the biblical record, towards a more esoteric view of our world.

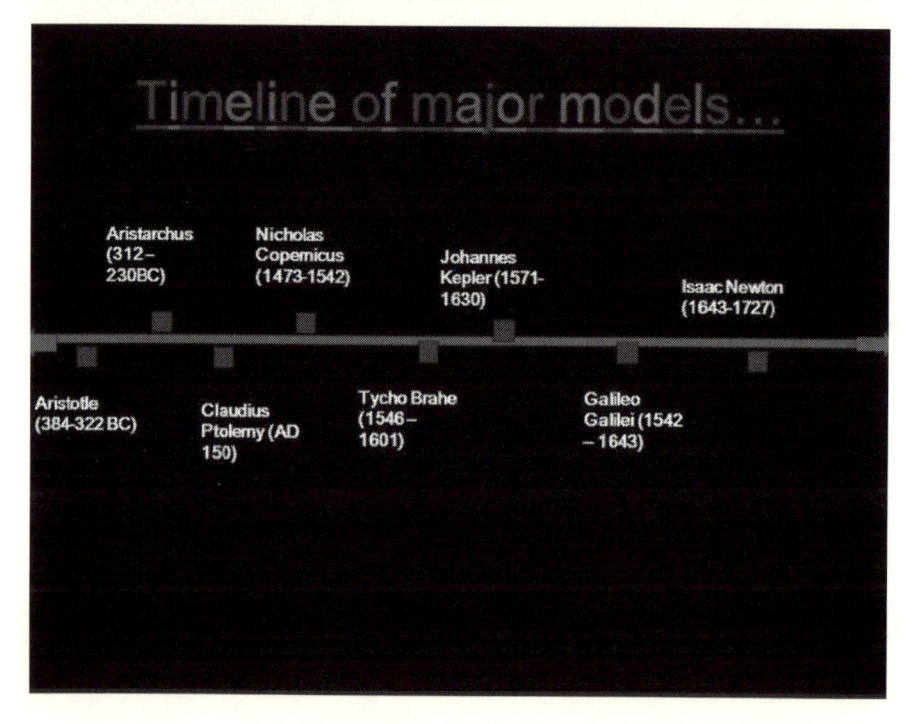

Timeline of major models...

- Aristarchus (312–230BC)
- Nicholas Copernicus (1473-1542)
- Johannes Kepler (1571-1630)
- Isaac Newton (1643-1727)
- Aristotle (384-322 BC)
- Claudius Ptolemy (AD 150)
- Tycho Brahe (1546–1601)
- Galileo Galilei (1542–1643)

- Pythagoras (570 – 495 BC) The idea of a spherical Earth was started by Pythagoras around 500 BC. As previously noted, Pythagoras speculated that the earth was round because he believed that roundness was the most perfect of geometrical shapes. Pythagoras did not experiment to determine the shape of the earth objectively. This was all determined though his best guess.

- Aristotle (384–322 BC) - Aristotle also assumed that the earth was round based on the following observations of the moon. According to Aristotle, the shadows and shapes on the moon were caused by the earth. As stated by him, "how else would eclipses of the moon show segments shaped as we see them? As it is, the shapes which the moon itself each month shows are of every kind." Aristotle also based his assumptions on the observation of the stars as well, to which he stated, "All of which goes to show not only that the earth is circular in shape, but also that it is a sphere of no great size: for otherwise the effect of so slight a change of place would not be so quickly apparent. Another guess. (Aristotle, "On the Heavens," Book II, Chapter 14, The Works of Aristotle, Oxford University Press; pp. 297-298.)

- Aristarchus of Samos (310 BC - 290 BC) was an ancient Greek mathematician and astronomer who came up with an alternative astronomical hypotheses tin order to address some of these concerns about a geocentric earth. Aristarchus was born about 20 centuries before Copernicus and Galileo but also claimed the sun, not the earth, was the fixed center of the universe, and that the earth, along with the rest of the planets, revolves around the sun. He also said that the stars were distant suns that remained unmoved, and that the size of the universe was much larger than his contemporaries believed (Ancient History Encyclopedia, 2019).

- Eratosthenes of Alexandria (276 BC-194 BC) is the Greek mathematician that was said to predict the size of the earth by merely observing how the sun rays fell upon the earth at different points and angles. Although his theory came under serious speculations, it triggered other scientists to take charge to find an accurate way to measure the size and distance across the earth.

- Claudius Ptolemaeus, (100 CE -.170 CE), Ptolemy argued, that the earth is a stationary sphere at the centre of a vastly larger celestial sphere that revolves at a perfectly uniform rate around Earth, carrying with it the stars, planets, Sun, and Moon—thereby causing their daily risings and settings. Nicolaus Copernicus , (1473-1543 CE), Polish astronomer who proposed that the planets have the Sun as the fixed point to which their motions are to be referred; that the Earth is a planet which, besides orbiting the Sun annually, also turns once daily on its own axis; and that very slow, long-term changes in the direction of this axis account for the precession of the equinoxes. This representation of the heavens is usually called the heliocentric, or "Sun-centered," system—derived from the Greek Helios, meaning "Sun."

- Tycho Brahe (1546, 1601 CE) Danish astronomer whose work in developing astronomical instruments and in measuring and fixing the positions of stars paved the way for future discoveries. His observations included a comprehensive study of the solar system and positions of more than 777 fixed stars. Tycho a self-acclaimed sorcerer, and astrologer designed the model of the universe that carved the way for the study of advanced astronomy (Britannica, 2019).

- Galileo Galilei, (1564-1642 CE), Italian natural philosopher, astronomer, and mathematician who made fundamental contributions to the science of motion, astronomy, and strength of materials and to the development of the scientific method. His formulation of (circular) inertia, the law of falling bodies, and parabolic trajectories marked the beginning of a fundamental change in the study of motion (Britannica, 2019). Modern day astronomers believe that it was his discoveries with the telescope that paved the way for the acceptance of the Copernican heliocentric system, but his advocacy of that system eventually resulted in an Inquisition process against him.

- Sir Isaac Newton (1642 – 1727), the noted English scientist and mathematician, wrote many works that would now be classified as occult studies.

These occult works explored chronology, alchemy, and Biblical interpretation (especially of the Apocalypse). Newton's scientific work may have been of lesser personal importance to him, as he placed emphasis on rediscovering the occult wisdom of the ancients. In this sense, some have commented that the common reference a "Newtonian Worldview" as being purely mechanistic is somewhat inaccurate.

After purchasing and studying Newton's alchemical works in 1942, economist John Maynard Keynes, for example, opined that "Newton was not the first of the age of reason, he was the last of the magicians" (Wiki, 2019).

II. Hermes Trismegistus a leading man in our current Global worldview.

"The name Hermes Trismegistus ("Thrice-Great Hermes") is a syncretism of the Greek god Hermes and the Egyptian god Thoth." Hermes Trismegistus is the founder of the collection of religio-philosophical (occult) texts originating between 100 and 500 A.D. in Egypt that became known as the Hermetica.

The Hermetica is a category of popular late (Roman) antiquity literature purporting to contain secret wisdom. Compiled from a more extensive literature by Italian scholars (i.e., Ficino) during the Renaissance, it became the Corpus Hermeticum. Hermes Trismegistus ("thrice-great Hermes") secret magic and arts were embraced by many founding philosophers. In addition, his teachings have permeated nearly every culture worldwide.

In Hellenistic Egypt, the god Hermes was given as epithet the Greek name of Thoth. Both Thoth and Hermes were gods of writing and of magic in their respective cultures. Thus, the Greek god of interpretive communication was combined with the Egyptian god of wisdom as a patron of astrology and alchemy. In addition, both gods were psychopomps; guiding souls to the afterlife. Hermes Trismegistus might also be the man who was the son of the god, and in the Kabbalistic tradition that was inherited by the Renaissance, it could be imagined that such a personage had been contemporary with Moses, communicating to a line of adepts a parallel wisdom, from Zoroaster to Plato.

Hermes Trismegistus is often pictured holding a globe. There is no doubt that he was one of the most influential men who guided our current global worldview. He was considered a great teacher - better known as the Thrice Great Hermes of whom Albert Pike makes a parallel to Grand Master Hiram in his third-degree monograph (Morals and Dogma, Pg. 78).

III. The Occult Backgrounds of the Men who founded the Global Worldview.

Most people today are familiar with the quote from Arthur C. Clarke, his third self-described "law" which says: "Any sufficiently advanced technology is indistinguishable from magic" While this might seem like an old adage, it actually is quite true for many men of science. Isaac Newton himself was referred to as the last magician. Most people would not think of men like Tycho Brahe, Nicolaus Copernicus, or Isaac Newton as men of magic; however, this is exactly how they saw the world of their time. Anyhow, my point is that the more we continue digging into the development of modern science or so-called science, with its rather glaringly false foundational and philosophical assumptions, the more we can see that they are based on magical principles. How did this happen?

Isaac Newton: The Last Magician

In the Western World, many have grown up in a culture and an educational system which for the most part, has regarded the development of scientific knowledge as largely the product of some vague concept of the Judeo/Christian belief system. This generic idea that it was the Christianized nations of Europe, who because of their Theistic foundation, were leaders in the empirical study of science without, borrowing from non-rational ideologies, is only a myth. We have been taught in school that true science is making predictions about phenomena that can be predicted, gathering, measured, weighed, and tested.

Christians believe and were taught that a rationale Creator made the world in an orderly way, that could be studied and understood. For years, we have taken comfort in the belief that our system of scientific study and development is based on facts rather than fables. And, that is was facts that guided the discoveries of men like: Isaac Newton, Galileo, Benjamin Franklin, Thomas Edison, Albert Einstein, etc. All geniuses that we have heard about in school. However, this worldview does not hold up well, when we examine our history. Not surprisingly, it seems to be the Christian culture today that is the most blind in this matter, preferring to cling to this mythology about the Christian roots of Science. This might be because it plays such a large role in our current approach to Christian apologetics. It turns out that many Christian organizations use science as a way to explain the creation. As a matter of fact, many Christians have turned to sciences in hopes of vindicating the Genesis account in opposition to the Darwinian paradigm.

But when we listen to secular historians expound upon things that pertain to the foundations of many scientific theories, we find that the development of many theories of science, are not so cut and dry. As a matter of fact, if we are willing to lay down a lot of those rosy-colored, Christianized views of our scientific history in Europe, America, and throughout the West, we can plainly see the esoteric and even magical underpinnings of many very foundational theories. Both Alchemy and the Kabbalah were used in the development of chemistry, atomic theory, gravity and even the development of Copernican astronomy. Moreover, all of our cosmology, traces back to the Occult, Mystery Schools. The very roots of science itself, are largely developed through the occult.

How Hermes in Europe

Hermetic philosophy came into Europe towards the beginning of what we now refer to as, the Renaissance. The figure of Hermes Trismegistus sheds an incredible amount of light as to how ancient esoteric teachings penetrated Europe and gave rise to the period known as the Renaissance, the Enlightenment, and finally the Scientific and Industrial revolutions. It is rather astounding to learn just how highly regarded the figure of Hermes Trismegistus to the European Renaissance mind was. He was considered by scholars from Augustine to Thomas Aquinas, to be not just a literal figure of history, but a pagan Egyptian priest, who, somehow, apparently because he was believed to be more or less a contemporary of

Moses, was viewed as sort of this quasi-patriarchal figure. They really elevated him essentially to the same level as Moses or Abraham, but as this kind of extra-biblical prophet, who some have even argued gave credence to many Christian doctrines.

Hermes was regarded by many as the quintessential magus, (a priest). However, person of antiquity was no more than a pagan magician. It is still interesting, though, that this magician somehow came to receive this incredibly favorable treatment by European Christian theologians and scholars, alongside Europeans who were deeply attracted to the Occult elements of this tradition. And so in the 15th century, we have the Italian scholar, a Catholic Priest, Marsilio Ficino, who had been commissioned by the wealthy Medici family to translate the works of Plato, and yet when the Medici suddenly came into the possession of a Greek manuscript said to be the writings of the magnificent Hermes, Ficino was ordered to halt all work on Plato and translate the Hermetic text with the utmost priority.

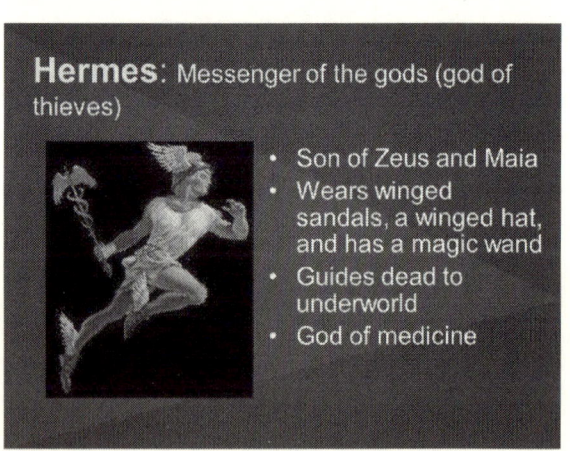

Hermes: Messenger of the gods (god of thieves)
- Son of Zeus and Maia
- Wears winged sandals, a winged hat, and has a magic wand
- Guides dead to underworld
- God of medicine

What resulted was the work known thereafter as the Corpus Hermeticum, and this volume, along with other works such as the Asclepius, began to be spread around Europe among scholars and

Renaissance thinkers, being regarded as quite harmonious with the increasingly popular Kabbalistic writings, (and rightfully so, since they are derived from the same Occult origins) and all of which quite plainly was fairly inseparable from the spread of Copernicanism in Europe. While the Corpus Hermeticum contained a fair amount of more philosophical/ideological content, it also contained descriptions of outright occult rituals, such as how to summon astral energies down from above into stone idols and animate them, and so is quite deserving of the categorization as witchcraft.

Helios
School of Esoteric Science

Helios is one of many Mystery Schools that teach how to use magic.

IV. The Men who were most influenced by the Hermetic Magic (Witchcraft).

One prominent figure within the rise of Copernicanism whose Hermetic influences have been thoroughly documented is the well-traveled figure of Giordano Bruno, who first proposed that stars are distant suns, and was killed by being burned at the stake. Frances Yates, a 20th century historian who specialized in Renaissance history, wrote an entire book on this titled "Giordano Bruno and the Hermetic Tradition", which you can now read online for free.

Copernicus himself referred to Hermes in this quote about the sun:

"In the middle of all sits the Sun enthroned. In this most beautiful temple could we place this luminary in any better position from which he can illuminate the whole at once? He is rightly called the Lamp, the Mind, the Ruler of the Universe; Hermes Trismegistus names him the Visible God, Sophocles' Electra calls him the All-seeing. So the Sun sits as upon a royal throne ruling his children the planets which circle around him."

Isaac Newton is well known to have been an alchemist and student of Kabbalah, supposedly having endeavored to translate the Emerald Tablets of alchemy, but he too was a student of the Corpus Hermeticum. So, we have Nicolaus Copernicus, Johannes Kepler, Robert Boyle, Isaac Newton, all heavily influenced by Hermeticism, and of course, Francis Bacon as well, whose writings would seem to eventually inspire the creation of the first official institution of scientific study, the Royal Society.

Interestingly enough, there has been a fair amount of research done by numerous people examining the founding members of the Royal Society and their connections to both Freemasonry and Rosicrucianism.

Vincent Rhodes

Emerald Tablets

Graven Image

V. Gravity

It was Newton who developed the theory of gravity. Newton theories were heavily influenced by the writings of Hermes Trismegistus. Among Newton's notes were found the following quote in which he translated the writing of Hermes Trismegistus. In these notes, Newton wrote that the Sun is the Father of creation, the moon is the mother, and the earth is the nurse.

Hermes Trismegistus

It was Hermes Trismegistus spinning ball (spherical earth) that obviously led to the need for gravity. Without gravity, it would be impossible to explain a spherical earth in which people could walk on a spinning ball without falling off. Isaac Newton and Gravity Sir Isaac Newton believed that an invisible force called gravity that held planets in orbit around the Sun created action at a distance Newton hesitated to share his groundbreaking theory of gravity for fear that it would be branded as magic or witchcraft. Isaac Newton used hermetic magic in order to introduce gravity as a valuable force in the universe and as a true force. Isaac Newton and Witchcraft "I can make things move without touching them. I can make animals do what I want without training them. I can make bad things happen to people who are mean to me. I can make them hurt if I want to..." —Young Tom Riddle describes his magical skill, (in the famed Harry Potter series).

Many people believed Isaac Newton was a man of science; a scholar and a practitioner of mathematics. Textbooks tell us that he was a man of science; however, his use of Gnostic and occult philosophies, (such as the writings of Hermes Trismegistus) demonstrate that his teachings were rooted in idolatry and pagan beliefs. Actually, much of Newton's practices and beliefs were witchcraft. Isaac Newton's belief that invisible forces control the physical world, were very similar to the belief that the earth's elemental energy can produce changes in our physical world.

Graven Image

Witches believe that there are four elements in nature. The four elements are earth, air, fire, and water. According to most of their teachings, each of these elements has energy emanating from them that influence our physical reality. The problem with this view is it presupposes the existence of God but, in practice, denies his reality. There is some strength in this position: e.g. that unseen forces have access to our physical world. The problem is that it ignores the fact that Jesus exercises authority over both the spiritual and physical world. This practice and embracing the elements and elemental spirits is something that the Apostle Paul addressed in his letter to the Galatians: "Formerly, when you did not know God, you were in bondage to beings that by nature are no gods; but now that you have come to know God, or rather to be known by God, how can you turn back again to the weak and beggarly elemental spirits, whose slaves you want to be once more?" Even people of Paul's time believed that spiritual forces had authority over the physical world.

Newton's attempt to explain a force (gravity) emanating from nature, that could not be tested, weighed, or observed in the real world; and only could be observed from a spiritual vantage point, without acknowledging God, is blasphemy. This is why prominent astrophysicist, Neal deGrasse Tyson, has went of record to state that he does not know what gravity is. According to Tyson, gravity is a force that he cannot explain that governs the universe. However, Tyson's explanation excludes God as the force behind the action. This is absurd, of course, because science begins by recognizing that if you have a cause, there must be a causer. Believers know that the causer is God

This is why the writer stated that, "The secret things belong unto the LORD our God: but those things which are revealed belong unto us and to our children forever, that we may do all the words of this law" (Deuteronomy 29:29). A Historians View Newton & Demons According to, John Henry, a Fellow at the History of Medicine in London (1986), before Newton's time, consulting a demons was a means of taking a short-cut to the knowledge of natural magic. While scientist would never admit that they stand on the shoulders of occultist, Henry reveals that the secret powers at work in the natural world is really found through the conjuring of demons. Henry states that many philosophers of old recognized the validity of occult directed experiments. The fact is that this approach led to the phenomenal success of Isaac Newton's treatment of gravity. In response to G W Leibniz's disparagement of his concept of gravity as an occult force, Newton did not deny it.

"So history reveals that modern science was able to make such rapid gains in the 17th century only by plundering natural magic. Although Newton's work on gravity provides the most striking single example of the fruitfulness of notions of 'occult' qualities, many medical men contributed to this story" (Henry, 1986). A review of most scientific journals will credit Isaac Newton for coming to understand gravity by observing the falling motion of apples. According to an apocryphal meme, the apple hit him in the head; however, most scientists agree that this is a funny but untrue story. In truth, Newton needed a way to explain the motions of the moon. This explanation of the heavenly body's motion would be called gravity. Actually, gravity was necessary to deceive people into believing that our world is a round globe.

A 21 Century Graven Image Why the Globe is a Modern Form of Idolatry? Newton's theory sets the groundwork for what would later be one of the greatest deceptions in history! The Globe. As previously stated, without gravity, it would have been impossible to make people believe that they live on a globe.

However, the globe point of view, is by its very nature implicitly unbiblical and rooted in the doctrines (teaching) of demons.

VI. How It All Fits Together

Now, if we can fairly easily recognize Darwinian Evolution to be a repackaged Luciferian doctrine, masked in the guise of humanistic science, then when pondering the false cosmology of

Copernicanism. However, Copernicanism rest on the assumption of gravity. In order for the earth to be a globe in which we do not fall off, we must assume that there is an invisible force holding us fast to it. That force is gravity. Yet, gravity was honed out of the same Luciferian doctrines that crafted the heliocentric views of Copernicus and Darwinism.

Gravity" is also another one of these ancient mystical concepts that has been repackaged and re- introduced to the masses as "scientific fact". After all, "Gravity" is not merely taught to be the pulling force produced by humongous collections of mass, such as stars/planets/ moons etc., but it is also the very force which is alleged to have pulled all these celestial bodies together in the first place! According to the Big Bang, from the original "Singularity" sprang forth all the physical matter that is now still constantly being flung outward in every direction in an ever-expanding universe.

Francis Bacon, the famous occultist who is credited with basically being the father of the scientific method, wrote the well-known utopian novel "New Atlantis", which described the creation of an ideal college, dedicated to human discovery and knowledge, called "Solomon's House." The Royal Society was in many ways the blueprint for the Science departments for universities in the centuries that followed, and it is believed that this is the beginning of where the Occult roots and influences were successfully white-washed from the enterprise of science, allowing it to morph into the more contemporary concepts of Natural philosophy and scientific materialism we are familiar with today.

FRANCIS BACON

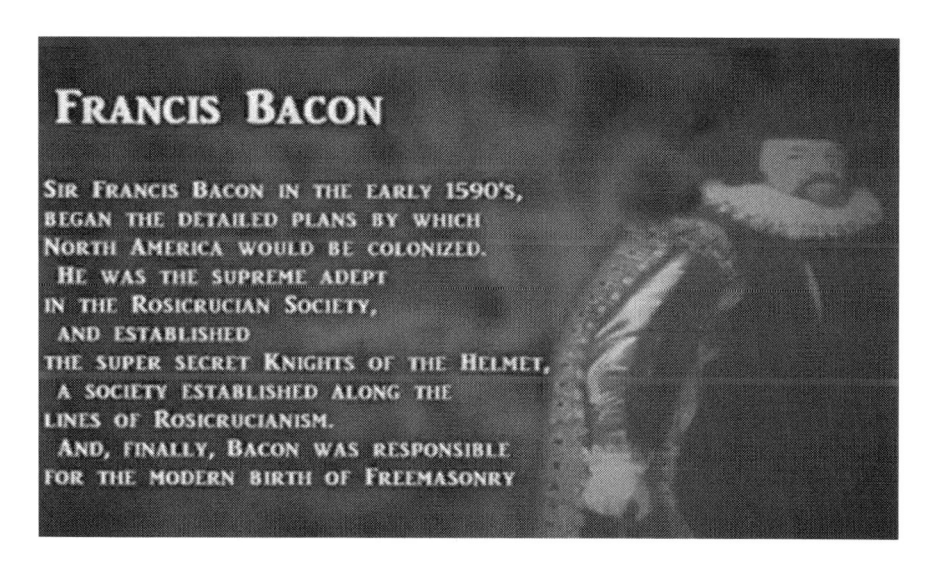

Sir Francis Bacon in the early 1590's, began the detailed plans by which North America would be colonized. He was the supreme adept in the Rosicrucian Society, and established the super secret Knights of the Helmet, a society established along the lines of Rosicrucianism. And, finally, Bacon was responsible for the modern birth of Freemasonry

Many people have equated Bacon's "New Atlantis" to the founding of the United States of America, with it's pervasive Freemasonic origins and all the esoteric symbolism of the streets of Washington DC, and so much more, and, I'd have to say that this theory probably has a lot of truth to it, but at the same time, is probably one a piece of the whole… If the building of "Solomon's House" was fundamental to the creation of that Luciferian utopia of the "New Atlantis", then, I now have to think that in all probability, this massive thing that we call "Scientism" is really almost another term for that same overarching thing.

The re-introduction and reestablishment, of those ancient "sacred sciences", or "seven liberal arts" of antiquity. In the days of the Early Christian church it was combatted as "Gnosticism", the heresy of the Mystery School teachings which was constantly trying to embed itself within the orthodox Christian gospel, whereby the acquisition of knowledge, or "Gnosis" was central to humanity learning to save itself. How is the current understanding of "Science" really any different?

The New Atlantis

Yates argues without the influx of Hermetic philosophy in the 15th to 17th centuries, there would quite simply have been no Scientific Revolution. Can you imagine that. According to Yates, the magical traditions of Hermeticism, which focuses on the influences of things on each other, on their interconnectedness in nature, and on the practices of observation of these influences, and the classification of phenomena and elements of nature, there's no such thing as science as we know it today.

Kabbalism, Hermeticism, (and also Neo-Platonism) hadn't entered in, then we're left with a somewhat uncomfortable question, as to whether or not the Bible on it's own would have ever inspired people to the level of scientific study and technological advancement that so rapidly occurred, between the rediscovery of all these ancient Occult teachings (i.e. the "Renaissance" or revival of said Occultism) and the modern day.

None to these men received their knowledge through the divine revelation of God. Instead, they boldly and broadly published their own musings from the hell bent halls of their own depraved minds. They could have simply relied on divine revelation, but could not because they were driven by another spirit. The notion that the world is a globe will definitely put this Bible verse in question, "the devil taketh him up into an exceeding high mountain, and sheweth him all the kingdoms of the world, and the glory of them..." Man can only understand this scripture through the divine revelation of God

The church will not want to associate with cosmology as it is mostly linked with activities like magic, the occult, witchcraft and ancient worship which is clearly not supported by the church. However, when we look clearly at the doctrine, it doesn't do justice to the belief in cosmology that attempts to explain some biblical verses and ideas, especially the globe shape that is not in agreement with the Bible.

CHAPTER 4
The Cult of the Globe

It cannot be stressed enough just how much the globe is part of a cult belief system born out of the Mystery Schools of a Gnostic belief system. Most people think that the globe (round Earth) is a product of pure scientific observation and discovery. They have no idea that the globe was first imagined in the minds of men without the benefit of the (so-called) NASA space organization. -A subject for another day.

Although the subject of satellites are in question, it should be understood that Gnostics announced the globe long before our modern times.

Gnosticism (from Ancient Greek: γνωστικός gnostikos , "having knowledge", from γνῶσις gnōsis , knowledge) is a modern name for a variety of ancient religious ideas and systems, that dates back to the ancient Mystery Schools of Egypt, Babylonian and Greek influence. These systems believed that the material world is created by an emanation or 'works' of a lower god (demiurge), trapping the divine spark within the human body. This divine spark could be liberated by gnosis, spiritual knowledge acquired through direct experience. Some of the core teachings include the following:

1. All matter is evil, and the non-material, spirit-realm is good.
2. There is an unknowable God, who gave rise to many lesser spirit beings called Aeons.
3. The creator of the (material) universe is not the supreme god, but an inferior spirit (the Demiurge).
4. Gnosticism does not deal with "sin," only ignorance.

5. To achieve salvation, one needs gnosis (knowledge).

The Gnostic ideas and systems flourished in the Mediterranean world in the second century AD, in conjunction with Middle Platonism (Wikipedia, 2019).

Pythagoras was a Gnostic. His views on secret knowledge are well documented by many of the secret organizations. One such secret organization are the Freemasons. To Albert Pike, the Grand Commander of the Freemasons, Pythagoras was a key figure among their constituency. Pythagoras is understood to be one of the most revered of ancient mystics and Mystery School teachers. Pythagoras was said to have been consecrated to the god Apollo even before his birth, Pythagoras was reared under the tutelage of teachers such as Thales and Anazimander at Miletus, but as a young man found himself unsatisfied by their seemingly disparate and contradictory forms of Gnosticism.

He set out to find a more "synthesized" universal Truth, and allegedly traveled through most of the great civilizations at the time, visiting the priests of various Mystery Schools in places like Egypt, Babylon and Chaldea. He basically combined all the Occult knowledge he could gather from the world at that time, and then eventually settled in Croton, Greece, where he started his own Mystery School, and developed a system of initiations and degrees, training through aestheticism, the majority of which was embedded within a complex code of numerical values and derivations. Thus, a vast amount of the concepts found in Occult Numerology and 'Sacred Geometry' about the shape and size of the earth, which have filtered down through the centuries to our own time can be traced back to Pythagoras himself.

The Doctrines of Devils that spawn the Globe

"Pythagoras refused the title of Sage, which means 'One who knows'. He invented and applied to himself that of Philosopher, signifying one who is fond of or studies things secret or occult. The astronomy of which he was taught was astrology: his science of numbers was based on Kabalistic principles. Everything is veiled in numbers." And truly, the further one might look into the beliefs and pursuits of individuals such as Pythagoras and Philolaus, the more it becomes plain that these were not simply individuals preoccupied with a rational and scientific analysis of the natural world, as our secular education system to often portrays them. Indeed, Pythagoras is revered by Plato and all the famous Greek Philosophers who came after him, and in that light we can understand just how pervasively Gnostic all of Greek "Philosophy" truly is, how it was seeded and guided not simply by the imaginations and musings of speculative men, but by the doctrines of Demons as they were passed down through Mystery tradition and ancient secret societies.

Is it any wonder that occult brotherhoods such as the Freemasons have such a high regard for Pythagoras? Is it at all surprising to discover that his mathematical codes and numerological interpretations of the universe such as the "tetractys" are interwoven throughout Kabbalistic and Masonic symbolism? Freemason Albert Pike says in "Morals and Dogma": And so from afar, when we step back and consider that neither Darwinian Evolution and Copernican Cosmology are not based on authentic scientific observation and experimentation, but rather on presupposed philosophical assumptions, it should also come as no real surprise that the ideological roots of both assumptions are practically the same. This only makes sense, really, because there is a rather obvious interdependence between the two.

Darwinian Evolution really can't even be conceived of outside of a massive, ever-expanding, self-creating Universe. A recognized geocentric cosmology would reveal the true absurdity of Evolution's basic premises. At the same time, the limitless expanse of the Copernican Cosmos are offered as the last great challenge for humans to conquer in their Evolution as a species, and so, their is an undeniable symbiotic relationship between these two assumptive maxims which we are all inculcated with from the earliest age. Both are designed to shape our perceptions of human origin, and human destiny alike.

II. The Scientific Merges with Occult Teachings

What is so interesting about looking at the example of Pythagoras, and the whole of 'Greek Philosophy' really, is that it could almost be said to be the first example of Esoteric/Occult knowledge being taken and formulated in such a way so that the "uninitiated" were still being given a version of the teaching on a level which was merely mechanical and nonspiritual. Greek Gnosticism is really where the ideological division between matter/spirit started, the false division between physical/ ethereal, mechanical/mythical, scientific/religious…

It's actually a basic tenet OF Gnosticism, which pits the material against the spiritual, rejecting the Biblical concept of hell/hades and replacing it with the physical reality itself. So indeed, Gnosticism contains an Esoteric and exoteric exposition for everything, which has more or less been the template for all Occulted information ever since.

GNOSTICISM
AND ITS INFLUENCE ON CHRISTIANITY

I believe this very much applies to our own time, where these Gnostic doctrines have been quite meticulously repackaged into forms which genuinely believe themselves to be wholly materialistic, empirical and scientific. So many adherents of Evolution today are staunchly materialistic in their worldview, completely oblivious to the fact that their beliefs stem from intensely religious and mythological beginnings. Currently we can see philosophies at the core of Evolutionary theory propelling the agendas of things like Transhumanism and the constant re-brandings of the New Age movement. The teachings of the Occult realm have always sought to pervert the Created order of the world and obscure the true identity of its Creator. As Blavatsky said: "It is from this sobering perspective that I am unable to sympathize with the notion that the Flat Earth debate is a "distraction". (Stranger than Fiction, 2015)

I concede that indeed it could be a distraction, from the Gospel of Salvation if it is not tied to it. However, if one investigates, it can be proven that the Flat Earth claims are much more important to the Christian life than one might realize. For one, the theory of

Graven Image

Evolution and Flat Earth are diametrically opposed. The proposition that God is right, the earth is flat, fixed and immovable, makes Evolution an unscientific satanic lie. One must then conclude that the purpose of Evolution is to prevent humanity from believing the Gospel and being saved.

The question of the globe, and the Cosmology of the universe as a whole, is no different, because it cannot hardly be separated from the other question. It is all, quite plainly, surrounding the question of origins, and destiny. In terms of origin, if it is true that men came from apes, then there is no need to be saved, since men were not created in the image of God. On the other hand, if men were created in the image of God and are accountable to him for how they have lived on this earth, then evolution is an evil designed to keep us from knowing God.

The "flat vs. globe earth" debate leans heavily towards the theological implications of the debate of End Time prophecy and mass-deception agendas. It is a prophecy that has kept me peering into this topic. For example, evolution ends with humanity evolving into transhumanistic beings, but the flat earth ends with the Lord Jesus returning to a flat earth from which every eye shall see him.

The whole driving force behind debate of the end times is science and its oppositions. Many people have seen the Jetsons and Little Einstein cartoons in which life is purely man without God. It is a world dominated and dictated by science. As a matter of fact, television as a whole is filled with images of galaxies and other worlds. This causes people to wonder about space and space travel. Then too, NASA's CGI technology makes it almost impossible to disentangle so called real space exploration from popular entertainment media. In the 1960s an iconic space adventure called

Star Trek debut on television. The lead actor, William Shatner begin each episode, with the words "Space, the final frontier…" People were hooked on the prospect that Jesus would not return to earth, but that we would evolve and travel the universe.

Mankind traveling through space is considered one of the most essential doctrines of devils. Since space travel involves absolute rebellion against God. The flat earth dictates that we stay where God has placed us. It informs us that God has placed us within an environment - a habitat that we cannot leave.

According to Genesis 1:6-7, God constructed a firmament. This firmament is hard and fast. We cannot leave it. This is why space travel is so important to the atheist and unruly. They simply can't accept the idea of there being a barrier. They can't accept the idea of a prison Earth. Never mind the fact that if they had never stumbled across the Flat Earth idea in the first place, they would've never in their whole lives worried about trying to travel to Antarctica. However once it is suggested that they can't go outside of a certain boundary, no matter how far away it is, staying put becomes intolerable. I find it to be such a perfect example of a broader psychological reaction which really manifests itself in a variety of ways. The more I think about it, the more it's hard not to see how the doctrine of Evolution and "outer Space" are inextricably linked.

Chapter 5

The Rotation of the Earth and Worship of the Sun

The Imaginary Rotating Earth and it Relationship to Evolution.

According to the Bible, the earth does not move. 1 Chronicles 16:30: "He has fixed the earth firm, immovable." Psalm 93:1: "Thou hast fixed the earth immovable and firm ..." Psalm 96:10: "He has fixed the earth firm, immovable ..."

Psalm 104:5: "Thou didst fix the earth on its foundation so that it never can be shaken." Isaiah 45:18: "...who made the earth and fashioned it, and himself fixed it fast..."

Even though the world of God states that the earth is fixed on foundations, scientists say that the earth is rotating. They say it is spinning at more than 1000 miles per hour at its equator. It might surprise you to hear, however, that scientists created this notion in order to support the theory of the Big Bang. The earth rotates presumably because of the Big Bang explosion.

"After the Big Bang only the three lightest elements existed, almost all Hydrogen, Helium and a pinch of Lithium. Eventually, and there are several hypotheses regarding galaxy formation, massive clouds of hydrogen came together under their own gravity, as they condensed, their mass became more centralized, making any movement of the gas speed up, like you do when you spin on a rotating chair and then tuck your legs in. The gas clouds rotated faster and faster, flattening out into disc shapes. This is why most large galaxies are flat. The first stars formed. When a star has thus used up all of its fuel, and is large enough, it goes supernova (smaller stars simply "puff" off their outermost layers). This scatters the new elements into space, and can produce even heavier elements than Iron.

Eventually, dust and rock in a particular orbit accreted into a large body. More and more matter would be "hoovered up" by its gravity over millions of years, and the energy of all the impacts and the heat from the gravitationally induced self-compression caused it to melt, whereby much of the heaviest elements sank to the center (Iron and Nickel) to form the core.

Another large, Mars-sized body had formed in the same orbit. Eventually the two collided. This added significant mass to the planet, and the material that did not return to Earth coalesced in orbit to form our moon.

Obviously, this story is in direct opposition to the word of God; however, willful scientist stubbornly contend with the word of God. This is obvious, since the word of God reveals that God created the heavens and the earth. Furthermore, there was no Big Bang that cause the earth, moon or other celestial bodies to rotate. A cursory observation of nature itself will inform us of this obvious fact. Even scientists have refuted the groundless theory of a rotating earth.

Real Science Refutes Earth's Rotation

Scientists who challenged the mainstream point of view was Michelson-Morley. Michelson-Morley experiment of 1887, at Chicago; the result of which might have undeceived even the most devoted believer in the theory of a spinning earth.

Graven Image

Professor Michelson was one of the physicists foremost in determining the Velocity of Light, while he has recently been described in the New York Times as America's greatest physicist; and it was he who— working in collaboration with Morley— in 1887 made the most painstaking experiments by means of rays of light for the purpose of testing, verifying, or proving by physical science, what really was the velocity of the earth. To express this more clearly. Astronomers have for a very long time stated that the earth travels round the sun with a speed of more than eighteen miles a second, or sixty-six thousand miles an hour. Without in any way seeking to deny this statement, but really believing it to be thereabouts correct, Michelson and Morley undertook their experiments in order to put it to a practical test; just in the same way as we might say.

"The greengrocer has sent us a sack of potatoes which is said to contain 112 pounds weight; we will weigh it ourselves to see if that is correct." More technically, the experiment was to test what was the velocity with which the earth moved in its orbit around the sun relative to the aether. "But to the experimenters' surprise no

83

difference was discernible. The experiment was tried through numerous angles, but the motion through the aether was NIL ! " Observe that the means employed represented the best that modern physical science could do to prove the movement of the earth through ethereal space, and the results showed that the earth did not move at all! " The motion through the aether was NIL ." . . . (Kings Dethroned, pg. 60, 1923).

The Solar System Is Occult

Before the Sun was formed, the watery Earth was here first. Therefore, the Sun could not have formed the Earth from rocks and star dust, as evolutionary theorist teach. Moreover, the Sun was founded on the fourth day of creation. According to the biblical record: "And God made two great lights; the greater light to rule the day, and the lesser light to rule the night: he made the stars also," (Genesis 1:16). Nowhere in scripture does God command the Earth to spin around the Sun. Otherwise the story of Genesis would have started with an introduction of the Sun. Today we have an entire school system that teach that we live within a solar system that none of us has ever seen.

The concept of the solar system first developed in ancient Egyptian. The Egyptians beliefs in Hermetic magic helped them to based their religion around the worship of the Sun. The Sun god rah rah was the god who was believed to live in the center of the universe. He was surrounded by eight other gods.

Those eight other gods were given names and attributes as deities. Today we know these gods as Mercury, Mars, Jupiter, Venus, Uranus, Saturn, Neptune and Pluto. According to the tradition of the Egyptians, Rai was the all-seeing - all-knowing Egyptian god who was worshiped as the creator man.

There Is No Solar System

We do live in a solar system with the Sun in the center of the universe. The earth is not the third planet from the Sun. This is part of the grand deception. A deception to hide the our Heavenly Fathers intentions from us when he created the Earth. The truth is that the Earth bears no resemblance to the planets or stars at all! The earth is a realm similar to heaven, except heaven is perfect. So, where did we get the idea that our earth is a spinning ball? The answer can be attributed to Nicholas Copernicus.

According to Copernicus mathematical calculations about its seasons drive the earth around the Sun. Copernicus distributed a handwritten book to his friends that set out his view of the universe. In it, he proposed that the center of the universe was not Earth, but that the sun lay near it. He also suggested that Earth's rotation accounted for the rise and setting of the sun, the movement of the stars, and the cycle of seasons was caused by Earth's revolutions around it. However, if we go back to scripture, it tells us that the Earth had no shape. Furthermore, the entire celestial realm was covered with water.

Copernicus did not come up with this belief based on mathematics and careful rationalism and scientific observation, but occult beliefs that are centered in an ancient pagan belief. According to Copernicus, the Sun is a majestic god whose rightful place is the center of the universe. According to Nicolas Copernicus: De Revolutionibus, I, 10. In the middle of all sits the Sun enthroned. In this most beautiful temple could we place this luminary in any better position from which he can illuminate the whole at once? He is rightly called the Lamp, the Mind, the Ruler of the Universe; Hermes Trismegistus names him the Visible God, Sophocles'

Electra calls him the All-seeing. So the Sun sits as upon a royal throne ruling his children the planets which circle round him.

After discovering this, it is obvious that Nicolas Copernicus held very none Christian beliefs. His assignment of deity to the Sun was nothing short of blasphemous. These were doctrines of devils of which the Apostle Paul warned the church about. Please notice what he says, "Now the Spirit expressly says that in latter times some will depart from the faith, giving heed to deceiving spirits and doctrines of demons" (1 Timothy 4:1).

Nicolas Copernicus took his foundational ideas for the Solar System from his pagan beliefs that the Sun is a god named Ra, whose rightful place should be in the middle of the universe.

Ra is the god of the occult and the underworld who is said to have created all things. His worshipers were often heard calling him the prince of eternity. The lord of life who feels the lungs with air and who gives breath to every nostril.

Please notice that the distinction between Ra and Jehovah is rooted in Sun worship. Our entire identity as a modern society, is rooted in the belief that the Sun is at the Center of our existence, rather than Jehovah-God. Most Christians (if not all) teach that the Sun is at the center of the -so called- solar system. I am astonished at the pervasive way that Sun worship has crept into our schools, movies, and worship. Most attend church on Sunday, without ever noticing that the days is called Sun-day.

Jesus recognized Satan's deception while here on earth. This is why he stated that "I am the way the truth and the life no man comes to the Father but through me (John 14:6). Nicolas Copernicus replaced the Son of God with the Sun of the universe for an entire generation. I imagine this is why he wanted to wait until he was dead before his work was published.

Satan's Number System Reveal Globe Hoax

As stated earlier, everything is veiled in numbers" The further one looks into the beliefs and pursuits of Pythagoras, the more it becomes plain that he was not an individual preoccupied with rational and scientific analysis of the natural world, (as our secular education system to often portrays him), Pythagoras was a member of a secret cabal of Gnostics who hid their true intentions behind numbers. Thus, Plato and all the famous Greek Philosophers came after him to be trained in ancient occult number magic.

This is how the doctrines of demons were passed down through mystery tradition and ancient secret societies. The further one looks into the beliefs and pursuits of Pythagoras, the more it becomes plain that he was not simply an individual preoccupied with rational and scientific analysis of the natural world, (as our secular education system to often portrays him), Pythagoras was a member of a secret Kabbalah of Gnostics. Indeed, Pythagoras is revered by Plato and all the famous Greek Philosophers who came after him, and in that light we can understand just how pervasively Gnostic all of Greek "Philosophy" truly is, how it was seeded and guided not simply by the imaginations and musings of speculative men, but by the doctrines of Demons as they were passed down through Mystery tradition and ancient secret societies.

Pythagoras view of the cosmos became widely accepted that the earth was a sphere in a universe which was itself also fully spherical. This claim was based on pure speculation and the belief that the circle or sphere was the most perfect of geometric shapes, and therefore appropriate for the shape of the earth. One of the so called proofs for this belief is found in the observation of a ship and its mast. As the vessel receded beyond the horizon, the ship's mast seem to sink over the curve of the earth. While we know this is not true today and merely matter of perspective, it serves it serves to demonstrate just how unscientific their observations were.

Moreover, since it was possible for the ancient Greeks to veil mysteries in numbers through trigonometric considerations, many people simply accepted that Greeks were wise beyond question. This is why Eratosthenes calculations concerning the circumference of the earth was never challenged. Since, at that time, no other assessment of the size and shape of the earth has been offered. It was believed by faith!

The Numbers of the Occult Earth

The earth is said to tilt on its axis at exactly 23.4 degree, on a 90 degree angle.

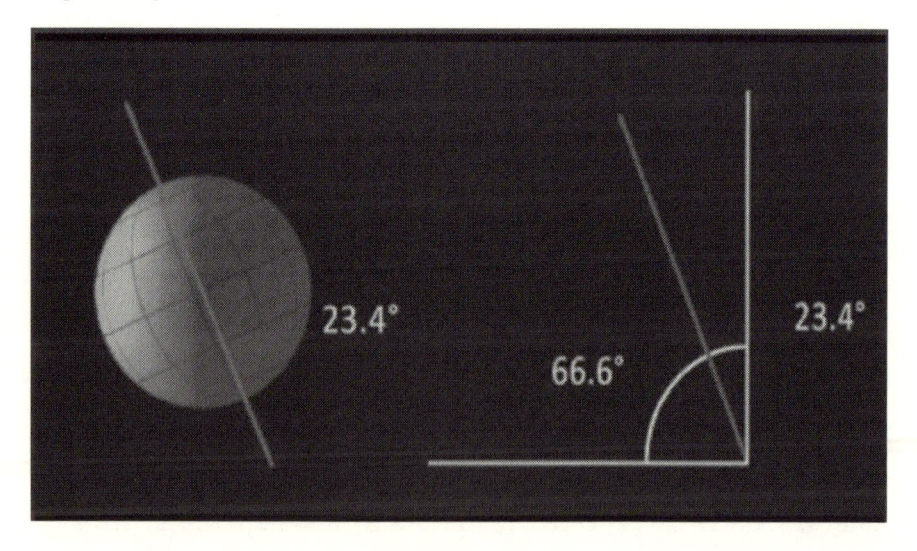

Here is wisdom. Let him that hath understanding count the number of the beast: for it is the number of a man; and his number is Revelation 13:18

The following is some numerology taken from spherical earth figures that is interesting.

"Speed of the globe's orbit: 66.600mph"

"Curvature in one mile squared: 666ft"

"The earth is on a 66.6 tilt."

"Earth's axis of rotation and its plane of orbit around the Sun: 66.6 degree"

"The polar circles are located near the poles of the Earth, at 66.6° N and S latitude."

Group Study

1. What does the Psalms says about the rotation of the earth?
2. Why does a rotating earth imply a Big Bang?
3. What Anti-Christian number are associated with the earth rotation on its (supposed) axis?

CHAPTER 6

How the Globe Supports A Fake Space Narrative?

Society's belief in the globe supports the false assumption that there is life on other planets. This is because many people view the earth as a planet. In actuality, the earth is not a planet. The word planet comes from the Greek word, planetes, meaning "wanderer." More accurately, it refers to the stars of the heavens. Right away we can see that the word does not fit the biblical description of the earth. This is because the earth is fixed and immovable (Psalms 104:5). The earth is unlike the surrounding constellations. Furthermore, there are no scriptures in the Bible that even remotely support the idea that planets were made for habitation.

Actually, the Bible does more to support the idea that planets are living beings more than anything else.

Even the book of Enoch supports the idea that planets are alive. Notice how Paul referred to planets in 1 Corinthians 15:40. Starting with verse 39, Paul speak about the human body as having flesh that differs from the bodies of the heavenly realm. Notice that he compares human bodies with animals and then stars: "Not all flesh is the same: Men have one kind of flesh, animals have another, birds another, and fish another. 40 There are also heavenly bodies and earthly bodies. But the splendor of the heavenly bodies is of one degree, and the splendor of the earthly bodies is of another." It is clear from this passage that Paul is referring to the stars as living beings which have a very bright and splendid appearance.

Such comparisons are all throughout scripture. Notice God question to Job, …6 On what were its foundations set, or who laid its cornerstone, 7 while the morning stars sang together, and all the sons of God shouted for joy? 8Who enclosed the sea behind doors when it burst forth from the womb,…(Job 38:7). Most people would say that this is only poetic, but it could be quite literal. Consider Jesus explanation of the mystery of the seven stars in the book of Revelation: "The seven stars are the angels of the seven churches..." (Revelation 1:20).

"Revelation 12:4 suggests an aspect of an angel's capacity. In biblical imagery, "stars" are symbols of angels. The verse implies that Satan coerced a third of these great beings to choose to submit to him and follow him in resisting God Himself as well as the outworking of His purpose in us. The Devil's persuasion and the angels' subsequent choices occurred in the distant past, and those who submitted to him are now demons against whom we wrestle (Ephesians 6:12). Stars, a symbol of angels, is used, meaning his angels—demons—were cast out with him. The Devil and his angels were cast to the earth. We have insight here into a major battle that took place in heaven, one that Satan and his angels lost, and they were cast to the earth. Unfortunately, that is where we live" (John W. Ritenbaugh, Sovereigty of God 2000).

In Genesis 1:14 we find this And God said, "Let there be lights in the expanse of the heavens to separate the day from the night. And let them be for signs and for seasons, and for days and years, 15 and let them be lights in the expanse of the heavens to give light upon the earth."

And it was so. 16 And God made the two great lights—the greater light to rule the day and the lesser light to rule the night—and the stars. 17 And God set them in the expanse of the heavens to give light on the earth, to rule over the day and over the night, and to separate the light from the darkness. And God saw that it was good. Genesis 2:1 Thus the heavens and the earth were finished, and all the host of them. In Job we find this 38:7 When the morning stars sang together, and all the sons of God shouted for joy? The Book of Deuteronomy tells us that God allotted the Nations to worship the *Host of Heaven* which it clearly identify as : The Sun Moon and Stars. Deuteronomy 4:19 And lest thou lift up thine eyes unto heaven, and when thou seest the sun, and the moon, and the stars, even all the host of heaven, shouldest be driven to worship them, and serve them, which the Lord thy God hath divided unto all nations under the whole heaven. We know for a fact that the Pagans did indeed worship the Host of Heaven including the Romans in the time of Jesus worshiping *Jupiter* and other planets as a gods.

Paul speaks about this of course and how through Christ all men have been freed from under the powers and can be transferred to Christs Kingdom, but at some point Christ will come back to Defeat the Kingdom of Darkness and put and end to it and all its people.

Admittedly, we cannot say for sure whether God is using metaphor; however, it is clear from scripture that God never intended for us to see stars (planets) as similar to the earth. Furthermore, the earth is not a star (planet) no more than a table is a lightbulb. Stars are to be seen as bright celestial beings whose duty is to serve God in the heavens.

Therefore, the belief that there is life on other planets comes from a total misunderstanding of what the planets (stars) actually are.

Is Heaven A Globe?

The consequences of thinking that earth is a globe are mind boggling for the Christian, to say the least. Because one cannot escape the misleading conclusions that such a belief will cause us to draw. For example, it is obvious from scripture that God created earth to be a copy of heaven. However, if earth a copy of heaven, then heaven must be a globe. Obviously, this is wrong thinking, because heaven is not a globe and most people never think of heaven as a globe, but a dwelling place. Yet, scripture clearly teaches us that earth is a shadow of what heaven is like (Colossians 2:17). God gave mankind dominion over the earth in the same way that he has dominion over heaven. In other words, he wanted human beings to imitate him in keeping the earth in order as he keeps heaven in order. The word dominion means the right to rule. Of course, mankind was never to rule independently of God's wishes. This is why God made human beings in his image and likeness. Sin marred mankind's resemblance to God, so Jesus came in the form of mankind to restore what humanity lost when sin entered the world bringing death. This is why Jesus instructed his disciples to pray, that the Father's will be done

(in)

earth as it is done (in)

heaven.

But the question remains, why should the earth realm be any different in form than the heavenly realm? Obviously, God's aim was to make earth a replica of heaven. Remember, he made mankind after his image and likeness. He also made earth to be a shadow of heaven. However, the truth is that scientists have all but erased the biblical narrative of our divine creation and order through evolution and cosmology that totally disputes the biblical record. Yet, most of what is expressed in the biblical record, fails to make sense unless it is viewed from the original interpretation of scripture. Such is the case with our understanding of the Lord's Prayer. When Jesus was on earth, he instructed his disciples to pray, "thy will be done (in) earth as it is (in) heaven." In other words, humans are (in) earth rather than (on) a globe. When we imagine heaven, we imagine a realm of existence that is inside and exclusive to outside intruders. Yet, we have no problem envisioning earth as a spinning ball in space with no boundaries.

Heaven has boundaries and gates, yet the scientific view of our earth supports no such boundaries. Scientists prefer us to believe that we live on a round ball with no boundaries. We can imagine that in heaven as a closed environment but not earth. This is why the false belief in Aliens from outer space is so prevalent among Christians. Yet, we know that heaven is exclusive and requires an invitation. The invitation is found in our acceptance of the shed blood of Jesus Christ as the full payment of our sins. Following this, Jesus bids us to come into heaven through these words: "In my Father's house, there are many mansions" (John 14:2).

Actually, the point of the Lord's Prayer is that we learn to submit to our Heavenly Father's perspective on things. The Father demands that we live within his will and under his supervision. We are to pray that his Kingdom comes to earth. That his will be done on earth as it is in heaven. Please notice that Jesus never made provision for space travel. He never said, thy will be done on Mars because space travel is not possible. Truthfully, we live in a closed system called earth.

The Church, the Globe, and the Alien Deception

Perhaps some of the most surprising support for life on other planets comes from Christian ranks. Pat Robinson, of the famed 700 club, has publicly spoken of his belief in evolution and the big bang.

Although he has not stated that he believes in life on other planets, he still cannot escape the deception of the globe which support the paradigm.

Pope Francis has spoken of life on other planets and even suggested that he looked forward to baptizing Aliens.

Another religious group that has thoroughly embraced the ball earth are the Mormons. No other church supports the idea of the globe earth more. I've always thought it odd that the LDS church claimed itself to be a Christian church when the majority of their doctrines were completely foreign to the mainstream doctrines of the Christian church established in the book of Acts.

They believe in a different ending other than the one described in the biblical text according to the Mormon's eschatology, the end of the age on earth, will be followed by space travel. Each good Mormon will inherit their own planet. Even the LDS founder, Joseph Smith, was visited by an extra terrestrial (angel) in 1830.

Smith who lived from (1805-1844) was selected and directed by the angel Moroni to translate these golden plates into English. Smith published what he said was an English translation of these plates, the Book of Mormon. According to Smith, Plates revealed another revelation for America. This new found faith would no longer center on the salvation of humanity, but the eternal progression of mankind.

Joseph Smith and Brigham Young, the second president of LDS, (1847- 1877) also changed the goal of the gospel of Jesus Christ, which is to save men from their sins, to another, which is to progress to become gods. Now this clearly goes against scripture, since scripture teaches that Jesus Christ is the final word given unto mankind; as Hebrews1:2 states (Berean Study Bible)

Previously, God spoke through various prophets and means, but "in these last days has spoken to us in His Son, whom He appointed heir of all things, and through whom He made the ages" A 21 Century Graven Image Why the Globe is a Modern Form of Idolatry? Scripture states: "But even if we or an angel from heaven should preach a gospel contrary to the one we preached to you, let him be under a divine curse!" (Gal. 1:8).

This is why accordingly, the LDS church's teachings are considered accursed, because they teach a different revelation of the gospel, other than found in the New Testament. Man will become a god and seed his own planet According to the Mormon faith; the earth age will end with Man becoming a god and seeding his own planet. The belief that man can become a god is only half of the belief. The other half is that our God was once a man.

The doctrine that humans can progress to exaltation and godhood is commonly taught within the LDS Church. Lorenzo Snow, the Church's fifth President, coined a well-known couplet: "As man now is, God once was: As God now is, man may be." So, the LDS Church teaches Man will have his own planet and enter the next age sealed to eternal spouses. This means that the man can have as many wives as needed to repopulate the planet. This might explain why polygamy is most popular among the church of Mormon

There is a clear contradiction between LDS teaching and the Bible. Since it teaches there will not be marriage in eternity. Jesus said that in eternity, there will be no marriage, or one given in marriage. But people will be like the angels Matt 22:30.

This view of the cosmos is quite inconsistent with scripture as well, since it does not fit the description that the word of God gives us of the world in which we live. According to scripture, God created the earth as a home for humanity. The purpose of mankind was to bring glory to God by subduing the earth. Gen. 1:28: "And God said to them 'Be fruitful and multiply, and fill the earth and subdue it; and have. dominion over the fish of the sea and over the birds of the air and over. Every living thing that moves upon the earth.

In this way, God gets glory by having mankind in his expressed image to rule over earth as he rules over heaven. . So according to Biblical cosmology, God is near. He sits in heaven which is right above the earth. God is not light years away. He is sitting above a three part system. Heaven, Earth, and Hell. It is he that sits upon the circle of the earth, and its inhabitants are as grasshoppers; that stretches out the heavens as a curtain and spreads them out as a tent to dwell in: (Is. 40:22).

So we see God sitting on a throne, above the circle, not a sphere, which is a different Hebrew word, but a flat sphere, above which are the waters. Notice that God sees all of the inhabitants of the earth like grasshoppers. Indicating that God is keeping watch over the inhabitants of the earth. Under the biblical model of the earth, planets do not exist, and space travel is impossible because, we, his creation are enclosed in a firmament that cannot be penetrated. According to the Bible, the firmament separates the waters above the earth from the waters below. Furthermore, the sun, the moon and the stars are enclosed within the firmament.

The Globe, Space Travel, and its Anti-Christian Agenda

Scripture reminds us that Satan is the father of lies. One of the massive lies that he has got the whole world to believe is that the earth is a globe. The globe means that our earth is a ball; a product of a Big Bang (explosion). According to most scientist, the reason it is spinning is because it was set in motion by the forces of gravity that gave rise to the big explosion. Scientist say that there is nothing special about our world. According to scientist, our world is just one out of a vast number of worlds surrounding us. Scientists use telescopes and computer-generated images to promote their anti-Christian teaching. They tell us that we are on a round ball similar to the surrounding stars and heavenly bodies. This is not true, of course, it fits perfectly with their anti-God, anti-Christian perspective. You might ask, why would they do this? One reason is because evil men have had a long history of using science to oppose the gospel. Notice what the Apostle Paul advised his young protege, Timothy, to do: "avoiding profane and vain babblings, and oppositions of science falsely so called." The reason their oppositions were regarded was because they were scientific. However, Paul warned Timothy not to listen to their vain babblings because its a false science. This same strategy of using science to manipulate the minds of men still is being perpetrated by the scientific community today and is orchestrated by the spirit of the anti-Christ.

Why Space Travel Is an Anti-Christian Concept?

According to the Apostle John, the anti-Christ is anyone who denies that Christ has come in the flesh (1 John 2:22). This also can apply to those who deny that he is yet coming again. It is crucially important for you to grasp the deceptive the doctrine of space travel. Space travel contradicts the words of the Lord Jesus Christ, because if it is true that people can ascend into the heavens by way of rockets, then Jesus would be a liar. Jesus clearly taught that heaven was above the earth. For example, Jesus said, "no man hath ascended up to heaven, but he that came down from heaven, even the Son of man which is in heaven" (John 3:13).

This word, "ascend," is the same word that is used for the ascension that Jesus performed in Acts 1:9-

12. In Acts 1:9-12, the Lord Jesus ascended into the heavens. The direction that he traveled was up. The angels stated that Jesus would return in the same manner is which left. The false notion that leads us to believe that mankind can leave the earth and travel through the void of space, contradicts validity of the gospel account. If it is true that people can leave the earth and travel through the heavens at will, then the words of Jesus cannot be true. The idea that earth is just a ball spinning in the void of space, seriously undermines the gospel account of salvation, the promise of heaven, and the age to come. It also discredits our creator's account of his design and purpose for him creating the earth. To fully understand just how deliberately the scientific community, push the Anti-Christ agenda, consider this, no scientists teaching in any major secular university may ever teach intelligent design as a viable option for our origin without the certain loss of his employment.

World leaders work diligently to discredit God through their massive Anti-Christian UN summits and their fake space programs. They force government schools to teach about space, planets, and the spherical earth. They mock and censor Christian thought and discourse. Every leader in the world is in on the ruse. They conspire to keep the secret that the earth is not a planet and that we live under the feet of God. Psalms 2 informs us of their evil agenda, "Why do the heathen rage, and the people imagine a vain thing? 2 The kings of the earth set themselves, and the rulers take counsel (conspire) together, against the Lord, and against his anointed..."

NASA

One of the leading organizations that promote the anti-Christ agenda is NASA. There is actually an agenda of NASA, to hide the truth of the flat earth. NASA helps to support Satan's lies by promoting an anti-biblical view of earth. They hide the fact that the earth is under a dome. Through their space program. they promote the big bang theory, evolution, and atheism.

NASAs own documents reveal that the earth is not a globe. According to their own flight manuals, they assume that the earth is flat and non-rotating in all of their calculations. (Derivation and Definition of a Linear Aircraft Model Pages: 6, 35, 55, 102 https://www.nasa.gov/centers/dryden/pdf/88104main_H-1391.pdf). All the governments on the face of the earth are controlled by Satan. He has deceived the entire world (Revelation 12:9).

Satan's plan is to replace Biblical prophecy with an alternative future for humanity. This future is one in which mankind saves themselves through their own devices. Under this narrative, mankind uses technology to create a future in which there is no judgment for sin. In this future, mankind merely evolves into a higher form of self.

However, according to the Bible, humanity final days on Earth are very near. Jesus reveals that all humanity will face the Judgment of

God. This is the reason that he came preaching that the Kingdom of Heaven is near (Matthew 4:17). According to scripture, this Kingdom will conclude with Jesus Christ reigning over humanity and the government of the world. The government shall be on his shoulder Isaiah 9:6). NASA's future does not include one in which Christ is reigning King.

They envision a future of space travel and conquest. They see earth as a launch pad from which they will eventually leave to travel to other worlds. NASA promotes this false doctrine. A doctrine that subverts the world of God, by teaching that human will escape the judgment of God through space travel. This will never happen because according to scripture, humanity will never leave the earth: According to God. 12And I saw the dead, great and small, standing before the throne. And there were open books, and one of them was the Book of Life. And the dead were judged according to their deeds, as recorded in the books. 13The sea gave up its dead, and Death and Hades gave up their dead, and each one was judged according to his deeds. 14Then Death and Hades were thrown into the lake of fire. This is the second death—the lake of fire.... (Revelation 20: 12-14).

In NASA's vision of the future, technology is the savior of humanity. Faith in God is replaced with faith in mankind. Who do you believe?

NASA, Technology, and the Anti-Christ

In Friedrich Nietzsche book called Antichrist (1895), Friedrich Nietzsche praised the Greek philosophers for the development of science. Nietzsche was a German philosopher, cultural critic. His most noteworthy manifesto was the work, "the death of God." Nietzsche used the phrase to express his idea that the Enlightenment had eliminated the possibility of the existence of

God. According to Friedrich Nietzsche, science was developed out of pure rationalism. This was a rationalism that excluded God and rejects the rule of Jesus Christ. Nietzsche thought that the advances of science by the Greeks was a great achievement because it killed off God. This is why Nietzsche wrote "God is Dead." To Nietzsche, "the Christian God is harmful and a crime against life. Nietzsche also believed that Christianity, is in opposition to reality, and "...mortally hostile to the 'wisdom of this world,' which means science" (Wikipedia, 2019).

In May of 2019 The Boston Pops honored NASA 50th anniversary since the fake moon landing. Conductor Keith Lockhart and his orchestra are conjuring the cosmos musical composures to celebrate the infamous Apollo 11 moon landing. Among the pieces composed was one in which Nietzsche was honored on behalf of Zarathustra," the 1896 Richard Strauss tone poem named after Friedrich Nietzsche novel.

NASA's 50 Years of Fraud

After 50 years, you would expect NASA to be a byword in human history. By now everyone should know that NASA was the space organization that tried something impossible but failed. However, as history would have it, NASA never ended. Instead, 50 years later

NASA is still hailed as one of the most important organizations in America. It is estimated that America's shell out more than 50 million dollars in tax revenue a day to this absolutely useless organization. What gives? Why is America so generous to this dinosaur of human failure. We never went to the moon; and furthermore, we stopped trying.

However, from the sounds of news bytes today, you would think that NASA is still alive and well. And they are.

As of today, July 16, 2019, newscasters around the world are still reporting the fake news that NASA landed on the moon more than 50 years ago. This was no small feat since nearly every newscaster in the world would need to be in on the ruse. What kind of synchronized effort does it take to get all news sources worldwide to report a fake story? Obviously, there are many newscasters who are more in love with money than the truth. However, the Strangerthanfiction channel on YouTube has compiled a handful of journalists who are willing to speak out against this fraud on the world.

Out of all the organizations that push the globe (schools, governments agencies movies, and the scientific community) no organization has been more convincing than NASA. NASA's 1969 photo of the earth was the first time that any picture was taken of the earth. It did not matter that the United States had claimed that they had a probe flight by Venus in 1962. It wasn't until the shot taken from the window of the space capsule, aboard the Apollo 11 space craft, that the world became fully convinced that Copernicus, Newton Galileo and all Greek scientists were right.

While this was a complete hoax, the vast number of Americans still don't know it. Only NASA and the government space agencies of the world, could claim to have sent men to see the globe (spherical earth). Yet, while the globe remains unproven, its existence is unquestioned. Practically everyone believes in the globe.

If you are new to the flat earth, I know what you are thinking. NASA proved that we live on a globe years ago. However, what you should understand is that NASA never went to the Moon and never will. NASA was founded in deception. If the kings of old, made god's for their people to worship, what makes us so sure that they would not do that today? The truth is, every image of the globe from space has been computer generated and artistically rendered. NASA Data designer, Robert Simmons, stated that the famous Blue Marble image of the earth is photoshopped. He said, it had to be.

According to NASA, the Blue Marble is an image of planet Earth made on December 7, 1972, by the crew of the Apollo 17 spacecraft at a distance of about 29,000 kilometers from the surface. It is one of the most reproduced images in human history. However, Robert Simmons was the designer of the photo. This means it is not an actual picture of the earth. A 21 Century Graven Image Why the Globe is a Modern Form of Idolatry? NASA Never Went To The Moon Although NASA claims that the picture of the earth was taken from space, by NASA's own admission, they have never left lower earth's orbit.

This means, our government lied about ever being able to leave the earth

"It's Photoshopped but it has to be"
Robert Simmons, NASA

Come on NASA* make your mind up, what does the earth really look like?

That means that the Apollo moon landings were a hoax. This is why they destroyed most of the thousand and thousands of reels of footage covering the moon landings.

However, the globe remains an unquestioned belief by practically every human being on the face of the earth. It is such an unquestioned part of the human reality that most people become violent when they are told that the globe is not real.

In summary, little has changed since the early days of old. Humanity still worships their idols that can neither see nor hear. (Deuteronomy 4:28 & Revelation 9:20). The Bible warns us that in the last days there will be strong delusion. 2 Thessalonians 2:11-12 King James Version (KJV) 11 And for this cause God shall send them strong delusion, that they should believe a lie: 12 That they all might be damned who believed not the truth, but had pleasure in unrighteousness.

There is no stronger delusion than the globe lie. Perhaps this is while they push it so hard. Because the globe supports every other lie of science. Without the globe, the gods of evolution and atheism would come crashing down. People would see their true purpose and turn to Christ so that they would be saved.

Made in the USA
Columbia, SC
21 December 2023